Handbook of Green Computing and Blockchain Technologies

Green Engineering and Technology: Concepts and Applications

Series Editors: Brujo Kishore Mishra, GIET University, India, and Raghvendra Kumar, LNCT College, India

Environment is an important issue these days for the whole world. Different strategies and technologies are used to save the environment. Technology is the application of knowledge to practical requirements. Green technologies encompass various aspects of technology which help us reduce the human impact on the environment and create ways of sustainable development. This book series will illuminate green technology in different ways, aspects, and methods. This technology helps people to understand the use of different resources to fulfill needs and demands. Some points will be discussed as the combination of involuntary approaches and government incentives, and a comprehensive regulatory framework will encourage the diffusion of green technology. Least developed countries, and developing states of small islands require unique support and measures to promote green technologies.

Handbook of Sustainable Development through Green Engineering and Technology
Edited by Vikram Bali, Rajni Mohana, Ahmed Elngar, Sunil Kumar Chawla, and Gurpreet Singh

Integrating Deep Learning Algorithms to Overcome Challenges in Big Data Analytics
Edited by R. Sujatha, S.L. Aarthy, and R. Vettriselvan

Big Data Analysis for Green Computing
Concepts and Applications
Edited by Rohit Sharma, Dilip Kumar Sharma, Dhowmya Bhatt, and Binh Thai Pham

Green Computing in Network Security
Energy Efficient Solutions for Business and Home
Edited by Deepak Kumar Sharma, Koyel Datta Gupta, and Rinky Dwivedi

Handbook of Green Computing and Blockchain Technologies
Edited by Kavita Saini and Manju Khari

For more information about this series, please visit: https://www.routledge.com/ Green-Engineering-and-Technology-Concepts-and-Applications/book-series/ CRCGETCA

Handbook of Green Computing and Blockchain Technologies

Edited by
Kavita Saini
Manju Khari

CRC Press
Taylor & Francis Group
Boca Raton London New York

CRC Press is an imprint of the
Taylor & Francis Group, an **informa** business

First edition published 2022
by CRC Press
6000 Broken Sound Parkway NW, Suite 300, Boca Raton, FL 33487-2742

and by CRC Press
4 Park Square, Milton Park, Abingdon, Oxon, OX14 4RN

CRC Press is an imprint of Taylor & Francis Group, LLC

Library of Congress Cataloging-in-Publication Data
Names: Saini, Kavita, 1976- editor. | Khari, Manju, editor.
Title: Handbook of green computing and blockchain technologies / edited by
   Kavita Saini and Manju Khari.
Description: First edition. | Boca Raton : CRC Press, 2022. | Series: Green
   engineering and technology: concepts and applications | Includes
   bibliographical references and index. | Summary: "This practical
   handbook will provide those dealing with Blockchain and Green Computing
   a computational perspective to access the current state-of-the-art,
   identify challenges, and provide input to develop strategies for
   addressing challenges. The handbook will have a practical oriented
   approach, include solved examples, highlight standardization, industry
   bodies, and initiatives. Case studies are included for a deeper
   understanding of Blockchain and are related to real-time scenarios. The
   book will analyze the current research and development in Green
   Computing and Blockchain analytics, study the existing related standards
   and technologies, and provide results on implementation, challenges, and
   issues in today's society. Those that will benefit from this book
   include researchers, software developers, both undergraduate and
   postgraduate students in Information Technology, Computer Science,
   Industrial Systems, Manufacturing, Communications, and Electrical
   Engineering"-- Provided by publisher.
Identifiers: LCCN 2021036958 (print) | LCCN 2021036959 (ebook) | ISBN
   9780367620110 (hbk.) | ISBN 9780367620141 (pbk.) | ISBN 9781003107507 (ebk.)
Subjects: LCSH: Blockchains (Databases) | Computer networks--Environmental
   aspects. | Green technology.
Classification: LCC QA76.9.B56 H335 2022 (print) | LCC QA76.9.B56 (ebook)
   | DDC 005.74--dc23/eng/20211012
LC record available at https://lccn.loc.gov/2021036958
LC ebook record available at https://lccn.loc.gov/2021036959

ISBN: 978-0-367-62011-0 (hbk)
ISBN: 978-0-367-62014-1 (pbk)
ISBN: 978-1-003-10750-7 (ebk)

DOI: 10.1201/9781003107507

Typeset in Times
by SPi Technologies India Pvt Ltd (Straive)

Contents

Preface

Promoting the use of renewable energy resources and limiting the use of non-renewable sources have been prime subjects of concern for over a decade now in order to safeguard our planet. The goals of sustainable development are of paramount importance, and are imperative to attain to ensure that there are sufficient resources and a healthy planet for future generations. Many non-renewable energy resources are already employed to create the required power – for instance, wind energy and solar energy.

Similarly, in the extensively developing area of technology, such a reasonable use of resources is crucial. Green computing is one such norm that promotes using technology and computing in an eco-friendly manner. It involves using, creating, or manufacturing and disposing of the devices used for computing in such a way that it does not have any adverse ramifications on the environment. Through this book, an effort in support of the idea of promoting green computing and establishing the rationale for green computing is made.

Chapter 1
The first chapter talks about how technology has become so dominant in the present-day world with its influence in almost all areas of life. But the same technology also has adverse effects on the environment. It is because of this fact that researchers and governments are now looking for greener alternatives in computing, such as green computing. The chapter in its later parts discusses the advantages of cloud computing, etc. Also, the history and development of blockchain and its applications are talked about in the first chapter.

Chapter 2
The second chapter discuss how the Internet of Things (IoT) is employed in various sectors, such as automation and agriculture. The chapter primarily focuses on green IoT and the applications it possesses in the field of healthcare. A technology to track the intake of medicines for tuberculosis is proposed and discussed.

Chapter 3
The third chapter discusses the concepts of neural networks and blockchain. The chapter begins by introducing neural networks and how they are created and how blockchain is linked to them. Data propagation, which is the transfer of data and information back and forth along the neural network, is talked about in this chapter. Finally, a review of papers is carried out, using Python modules and libraries, and how neural networks and blockchain can be used together to tackle the challenges currently faced is discussed.

Chapter 4
The authors of the fourth chapter have presented a significant appraisal of blockchain technologies established in related works and consider their suggestions for administrative processes. The framework is designed in preparation for additional research

into the considerable advantages of blockchain applications in the digitization of government and the participation of blockchain structure and its utilizations to fulfill community need for public value.

Chapter 5

The fifth chapter provides a brief overview of convolutional neural networks – i.e., trained on the basis of image data, which is propagated in forward and backward propagation. The chapter concludes with a literature review of research on the use of convolutional neural networks and blockchain technology together with open challenges in the similar domain.

Chapter 6

The sixth chapter provides the current scenario of the Indian agriculture sector and proposes an architecture to transform this sector. The limitations and challenges faced during the implementation of blockchain technology in agriculture, along with its advantages, are discussed.

Chapter 7

The seventh chapter talks about how the lives of many people have been affected by COVID-19. Many businesses suffered the inevitable by having to incur heavy losses. The chapter talks about a proposed blockchain model that can aid in secure and far more efficient transactions in the supply chain which is the major part of large-scale businesses in the modern era. The benefits of blockchain, such as transparency and immutability, and also the challenges faced by blockchain, are discussed in detail in this chapter.

Chapter 8

The eighth chapter introduces microgrids used for blockchains or distributed ledgers as a new and useful technology for energy companies, as well as for startups, government entities, and academia. The authors state that blockchains promise to be highly transparent, impossible to falsify, and security-tamperproof, particularly when employed in tandem with smart contracts. Blockchain activities and projects work to inform readers of innovative, scholarly, and theoretical research in the energy sector.

Chapter 9

The ninth chapter represents blockchain technology as a route bridge to assist government in all its usual processes and decisions. Nowadays, people as well as all electro-mechanical devices are in some way connected to the internet toward sharing and receiving data efficiently. Network servers may act as authorities and help to enhance efficiency in workflow, control, and enforcement of government laws as amended from time to time. Also, blockchain-based distributed ledger is secure, readily auditable and immutable to keep track of government data where data is secure and unaltered to eliminate forgery for personal interest. Governance may establish streamlined procedures for various government activities through emerging technologies such as blockchain. It removes the involvement of unethical third parties and corruption from most of the government activities– namely, the election process, resource procurement, political mergers, funds allocation and distribution, responsibilities of various representatives (leaders) and government political employees, and government asset management, to name a few. It is expected that the operations of the government will be transparently and efficiently managed and controlled through this emerging technology. This chapter covers relevant aspects of blockchain toward suitable governance.

Chapter 10
In the tenth chapter an advanced research and analysis is introduced in the context of healthcare. The main objective is to get the information about existing systems and their limitations and to seek a solution involving blockchain technology in the healthcare sector. The analysis approach consists of the collection of data and its properties with a bibliometric review. Hence, blockchain is playing a vital role in developing a system in the context of healthcare which will distribute information about diseases and manage the health of patients by keeping all their records.

Chapter 11
The authors of the eleventh chapter talk about SRAM gadgets to show genuine dangers. Many edge gadgets can be forged or cloned; the graceful chain is inadequately secure. It is an unquestionable requirement to distinguish edge gadgets' sourcing extraordinarily and check their legitimacy intermittently at run-time. All the world coordinates blockchain innovation to validate asset compelled, minimal effort edge gadgets. SRAM-based devices use genuinely unclonable capacities (PUFs) to produce one of a kind "advanced fingerprint" (gadget IDs). The authors use a "locally permissioned" blockchain framework to validate edge gadgets for a safeguard inside and out methodology. Gadgets can be validated occasionally to forestall gadget cloning. Target situations can be enormous and have differed trust among clients and come up short on a particular border; this "nearby" blockchain procedure is consequently relevant, particularly since blockchains gain security after some time. This proposed strategy ensures against such shams in portable settings inside an IoT framework as well.

Chapter 12
Chapter 12 begins by giving the reader an introduction to convolutional neural networks, then takes one step ahead by introducing the terminologies related to convolutional neural networks and architecture of the convolutional neural network. Data propagation is the most important step in convolutional neural networks trained on the basis of image data, which is propagated in forward and backward propagation. The final step in full understanding of convolutional neural networks comes with the implementation of the neural network using Python libraries. The chapter concludes with a literature review of papers on the use of convolutional neural networks and blockchain technology together with open challenges in the field.

We hope this book will be widely read. If we are to avoid the blunders of the past, then we need to change direction and start benefiting from the knowledge base created by the researchers. We did not have this chance a decade ago. Now is the right time. Finally, we want to convey our thanks to the editors, publishers, and authors of all the chapters who have helped bring this book to life.

Acknowledgments

We would like to thank the contributors for their valuable contribution and expertise in the area of green computing and blockchain technology. This book wouldn't have been possible without their constant support whenever required, which went far and beyond support. Without them we would never have brought this book to existence. We would also like to thank the reviewers who played a very important role in the review process. Without their support, this book would not have become a reality.

We are grateful to our family members who stood with us throughout the arduous process of completing this book.

We acknowledge CRC Press for allowing us to contribute to the emerging computer science area of green computing and blockchain. We would also like to thank the managing editor of CRC Press for the wonderful editorial support and guidance.

We would like to acknowledge the support of all the people who were involved in this project directly or indirectly.

Kavita Saini
Galgotias University, India

Manju Khari
Jawaharlal Nehru University, India

Editors

Kavita Saini, PhD, is currently Professor at the School of Computing Science and Engineering, Galgotias University, Delhi NCR, India. She earned a PhD at Banasthali Vidyapeeth, Banasthali. She has 17 years of teaching and research experience supervising Master's and PhD scholars in emerging technologies.

She has published more than 35 research papers in national and international journals and conferences. She has published 14 authored books for UG and PG courses for a number of universities, including MD University, Rothak, and Punjab Technical University, Jallandhar, with national publishers. Kavita Saini has edited many books with international publishers, including IGI Global, CRC Press, IET Publisher, and Elsevier, and has published ten book chapters with international publishers.

She has also filed several patents. Kavita Saini has also delivered technical talks on Blockchain: An Emerging Technology, Web to Deep Web and other emerging areas, and has handled many special sessions in international conferences and special issues in international journals. Her research interests include web-based instructional systems, blockchain technology, industry 4.0, and cloud computing.

Manju Khari, PhD, is an Associate Professor at Jawaharlal Nehru University, New Delhi. Previously, she worked with Netaji Subhas University of Technology, East Campus, formerly Ambedkar Institute of Advanced Communication Technology and Research, Under Govt. of NCT Delhi. She earned a PhD in computer science and engineering at the National Institute of Technology Patna, and received a Master's degree in information security at Ambedkar Institute of Advanced Communication Technology and Research, affiliated with Guru Gobind Singh Indraprastha University, Delhi, India. She has published 80 papers in refereed national/international journals and conferences (viz. IEEE, ACM, Springer, Inderscience, and Elsevier), and 10 book chapters in books published by Springer, CRC Press, IGI Global, and Auerbach. She is also co-author of two books published by NCERT of XI and XII and co-editor of ten edited books. She has also organized five international conference sessions, three faculty development programs, one workshop, and one industrial meeting. She has delivered an expert talk, been a guest lecturer at an international conference, and a member of the reviewer/technical program committee at various international conferences. Besides this, she is associated with many international research organizations as associate editor/guest editor of Springer, Wiley, and Elsevier books, and a reviewer for various international journals.

Contributors

Basant Agarwal
Department of Computer Science and
   Engineering
Indian Institute of Information
   Technology, Kota
MNIT Campus
Jaipur, India

K.P. Arjun
School of Computing Science and
   Engineering
Galgotias University
Greater Noida, India

M. Arvindhan
School of Computing Science and
   Engineering
Galgotias University
Greater Noida, India

Nikita Bhatia
School of Computing Science and
   Engineering
Galgotias University
Greater Noida, India

Chitrangada Chaubey
Sanskriti University
Mathura, Uttar Pradesh, India

Sweta Chaudhary
Ambedkar Institute of Technology
New Delhi, India

Ameya Chawla
Computer Science and Engineering
Guru Tegh Bahadur Institute of
   Technology
Guru Gobind Singh Indraprastha
   University
New Delhi, India

A. Daniel
School of Computing Science and
   Engineering
Galgotias University
Greater Noida, India

Mohit Dayal
Dr. Akhilesh Das Gupta Institute of
   Technology Management
New Delhi, India

Rupesh Kumar Garg
Computer Science
Central University of Haryana
Haryana, India

Seema Garg
Department of Electronics
   Communication Engineering
Ajay Kumar Garg Engineering College
   (AKGEC)
Ghaziabad, Uttar Pradesh, India

Ruchi Gupta
Department of Information Technology
Ajay Kumar Garg Engineering College
   (AKGEC)
Ghaziabad, Uttar Pradesh, India

Himanshu Dubey
Ambedkar Institute of Advanced
   Communication Technology and
   Research
Guru Gobind Singh Indraprastha
   University
Delhi, India

Jagannath Jayanti
Computer Science and Engineering
Guru Gobind Singh Indraprastha
   University
New Delhi, India

Manju Khari
School of Computer and System
  Sciences
Jawaharlal Nehru University
New Delhi, India

K. Kumar
Ambedkar Institute of Advanced
  Communication Technology and
  Research
Guru Gobind Singh Indraprastha
  University
Delhi, India

M. Kumar
Ambedkar Institute of Advanced
  Communication Technology and
  Research
Guru Gobind Singh Indraprastha
  University
Delhi, India

Sushant Kumar
Department of Computer Science and
  Engineering
SKIT
Jaipur, India

Mayur Mathur
School of Computing Science and
  Engineering
Galgotias University
Greater Noida, India

Swati Nigam
Department of Computer Science
Faculty of Mathematics and Computing
Banasthali Vidyapith
Banasthali, India

Kapil Parihar
Department of Forensic Science
Vivekananda Global University
Jaipur, Rajasthan, India

Nilanjana Pradhan
Department of Analytics
Pune Institute of Business Management
Pune, India

Shrddha Sagar
School of Computing Science and
  Engineering
Galgotias University
Greater Noida, India

Kavita Saini
School of Computing Science and
  Engineering
Galgotias University
Greater Noida, India

Mahipal Singh Sankhla
Department of Forensic Science
Vivekananda Global University,
  Jagatpura
Jaipur, Rajasthan, India

Mansi Saxena
Department of Forensic Science
Vivekananda Global University
Jaipur, Rajasthan, India

Jitendra Kumar Seth
Department of Information Technology
Krishna Institute of Engineering and
  Technology (KIET)
Ghaziabad, Uttar Pradesh, India

Anupama Sharma
Department of Information Technology
Ajay Kumar Garg Engineering College
  (AKGEC)
Ghaziabad, India

Anubhav Singh
Department of Forensic Science
School of Basic and Applied Sciences
Galgotias University
Greater Noida, India

Deepti Singh
Information Technology
ABES Institute of Technology
Ghaziabad, India

Swaroop S. Sonone
Department of Digital and Cyber
    Forensics
Government Institute of Forensic
    Science
Aurangabad, Maharashtra, India

Urvashi Sugandh
Department of Computer Science
Faculty of Mathematics and Computing
Banasthali Vidyapith
Banasthali, India

Anju P. Sureshbabu
TCS Blockchain Services
Infopark, Kochi
Kerala, India

M. Thirunavukarasan
School of Computing Science and
    Engineering
Galgotias University
Greater Noida, India

Vinita Tiwari
Department of Electronics and
    Communication Engineering
Indian Institute of Information
    Technology, Kota
MNIT Campus
Jaipur, India

Saif Ullah
School of Computing Science and
    Engineering
Galgotias University
Greater Noida, India

1 Green Computing and Blockchain Fundamentals

Saif Ullah, K.P. Arjun, and Anju P. Sureshbabu

CONTENTS

DOI: 10.1201/9781003107507-1

1.1　INTRODUCTION TO GREEN COMPUTING

Green computing consists of two words: 'GREEN' and 'COMPUTING.' GREEN concerns the environment, and COMPUTING means processing. So, it's computing in a way that doesn't harm the environment. Green Computing is often known as green technology which is the utilization of computers and their resources in an eco-friendly manner [1–3].

It's the study of designing, manufacturing, utilization, and dumping of computing devices with no or minimal impact on the environment. Computers are everywhere, so the waste generated from computers and their resources is significant and is directly harming the environment in the form of e-waste.

A lot of natural resources are consumed in using computer resources for manufacturing raw material, then to run the computer electronic power is required and for dumping again natural resources are exploited; all this badly impacts the environment [4, 5]. The objective of green computing is to limit the use of hazardous material, maximize energy efficiency, recycle e-waste, and promote biodegradable products, as shown in Figure 1.1.

1.1.1　HISTORY OF GREEN COMPUTING

In the 1990s computers were huge and so was their energy consumption, and they were not switched off when they were idle, so the waste generated from them was tremendous. According to the US Environmental Protection Agency (EPA), "in 2006, IT industry consumed 1.5% of the total electricity used in the United States, which is roughly 61 billion kilowatt-hours of electricity." This equals a bill of approximately $4.5 billion. Servers and data centers installed by the US federal government are responsible for 6 billion kilowatt-hours of electricity consumed.

This was taken into consideration and the following measures were adopted for the reduction of waste generated from computers [6].

- The first measure to adopt was started in 1992, when the EPA formed its Energy Star Program.
- In 2006, more stringent requirements for computational efficiency were added.

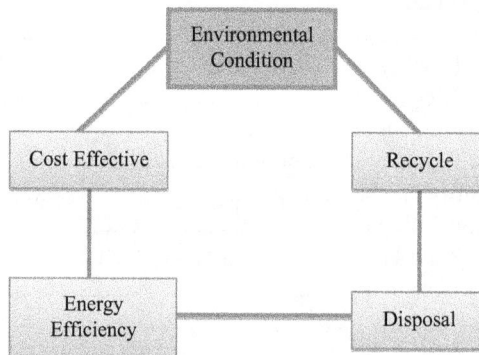

FIGURE 1.1　Objective of green computing.

- In 1997 UN's Kyoto protocol orders manufacturers to reduce carbon emissions and to calculate electricity utilization.
- The European Union in 2003 required Restriction of Hazardous Substances (RoHS) which limited the use of certain toxic substances to use in the manufacturing of electronic products.
- In 2006, the USA passed a law to promote and study computer servers that are energy efficient.
- In 2007, US President George W. Bush passed executive order 13423, which mandated all US federal agencies to utilize the Electronic Product Environment Assessment Tool (EPEAT), which rules standards to increase the efficiency and life of electronic products.
- TCO certification was launched by the Swedish organization TCO development which promotes the low electrical and magnetic emission from CRT-based computer displays [4].

1.2 ROLE OF GREEN COMPUTING IN INDUSTRY

ICT (Information and Communication Technology) is taking over all the world at a fast speed, along with bringing an IT revolution and alarming by the trouble it's causing the environment. We are all familiar with the rate of global warming and the depletion of the ozone layer; the level of CO_2 emissions caused by the IT industry is a significant percentage that disrupts the life cycle [7–9].

Along with initiatives from government, the industry is also responsible for adopting measures to promote green computing [1, 2]. The following are some industry measures for green computing:

1. On June 12, 2000, the Climate Saver Computer Initiative (CSCI) was started which focuses on reducing the electric power consumption of PCs in both active and inactive mode.
2. Green Grid was founded in 2007. This is a global-level trust which looks forward to advancing energy-efficient data centers and business computer ecosystems. Green Grid is backed by several companies in the industry, including AMD, APC, Dell, HP, IBM, Intel, Microsoft, Sun Microsystems, and VMware.
3. Companies in the IT industry are now using energy-efficient hardware to reduce carbon emissions.
4. Google is now using servlets and energy-efficient cooling which reduces the harmful impact on the environment.
5. To encourage companies to promote green computing there is an organization that facilitates them by giving special awards for implementing software and hardware which don't impact the environment.
6. IT companies are also responsible for disposing of e-waste responsibly and substances like aluminum, copper, iron, and plastic should be taken and reused, and non-usable substances should be recycled.
7. The industry should also focus on increasing the lifetime of its products so that they don't become non-operational after only a few years.

1.3 STRATEGIES FOR IMPLEMENTATION OF GREEN COMPUTING

As the IT industry bags for pollution as equal to the air industry, so controlling it is very important and certain approaches are listed down below [10].

1.3.1 VIRTUALIZATION

Virtualization is the creation of a virtual version of something, such as a server, a desktop, a storage device, an operating system, etc. It's a process that allows sharing a single example of a source or application between multiple clients or organizations [8]. It does so by providing a logical name for physical storage and by providing a reference to that source of material on demand. Figure 1.2 shows the varieties of virtualization that we used [11, 12].

The virtualization of the machine over existing software and hardware is known as Hardware Virtualization [3]. Virtual Machine provides a reasonable separate environment with substandard hardware.

1.3.2 ALGORITHMIC EFFICIENCY

Using an efficient algorithm will reduce the computer consumption in completing the computation function [4]. For example, substituting the fast search algorithm like hashed search algorithm with the slow search algorithm like linear will reduce the resource utilization and in a long run will reduce the wastage created by using the computer resources.

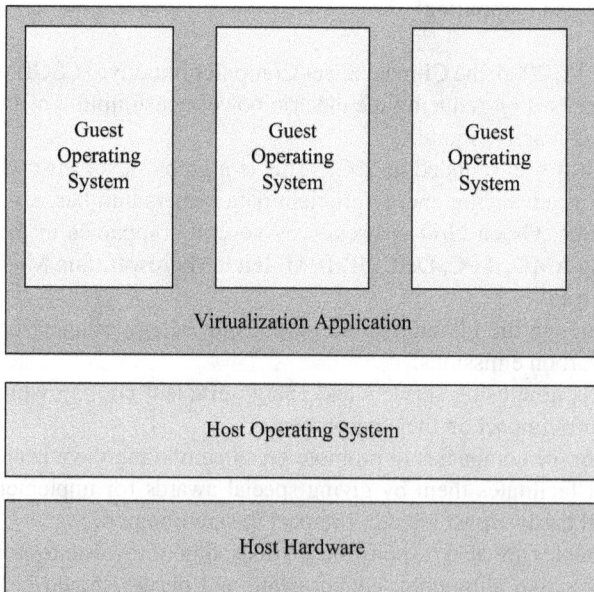

FIGURE 1.2   Different levels of virtualization.

1.3.3 POWER MANAGEMENT

The Advanced Configuration and Power Interface (ACPI) has set the standards to regulate the power-saving aspect, like allowing the system to turn off or to put devices like monitor and hard drives to sleep after a certain period of time. The system can also turn into hibernate mode when major parts are off, they also provide computer BIOS to manage the power functions. In addition to it, some systems allow the user to manually adjust the power loading according to their workload.

1.3.4 STORAGE DEVICES

In the earlier time, the large storage disk drive was used to contain the data, they used to be large in size and also have components in them but a now smaller disk or solid-state drive (SSD) stored data in flash memory which reduces a lot of extra wastage created by earlier large drives [7]. Now due to advancement in storage techniques, storage devices have moved to online which nullify the utilization of resources plus provide more space as compared to drives.

1.3.5 DISPLAY DEVICES

Light used in the display screen takes a lot of power consumption. Earlier CRT screens were used to consume energy more than the LCD screen monitor. LCD uses a fluorescent bulb while LED screens use light-emitting diodes in place of bulbs which cut the greater percentage of electric consumption [6]. Now, we are introduced to a feature Dark Mode, the main motive behind this mode is to reduce the display consumption to a greater extent, and it directly impacts the battery life growth and power consumption

1.3.6 CLOUD COMPUTING

The introduction of cloud computing has brought a great revolution for the green computing area. Bringing all the things to the cloud has put a full stop from the utilization of many resources, like bringing storage to the cloud has cut the consumption created from disk drives and online shopping platform has also created a hassle-free platform for the user and also saved a lot of liters of petrol on going to the market [10].

1.3.7 DATA CENTER POWER

Datacenter power is one of the most areas in the IT industry that consumes power consumption. Based on Greenpeace's study, data centers are accountable for 21% of IT industry electricity consumption which is a lump sum of 382 billion kilowatt-hours per year. So, techniques like virtualization are used to put a restriction on its electricity consumption [5]. The IT industry is also keeping a check on eliminating the underutilized data centers, Google is one of the best examples as it has been able to reduce 50% of electricity consumption of industry average [11].

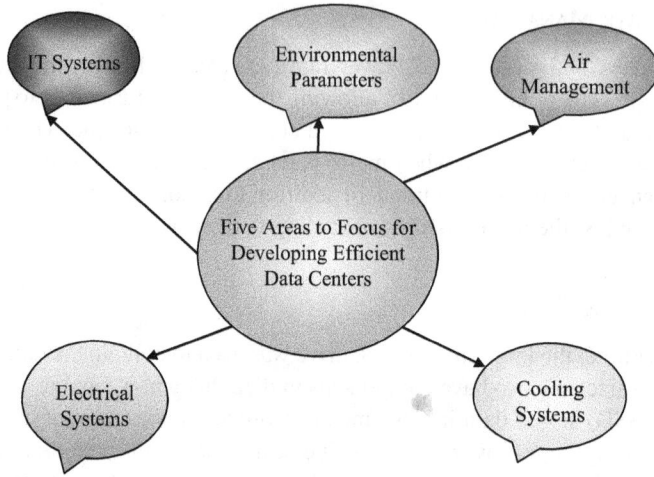

FIGURE 1.3　Five different data center design constraints.

1.3.8 DATA CENTER DESIGN

Since a data center accounts for most energy consumption of around 1.1%–1.5% of the world's total energy consumption in 2010, so the designing of data centers in a manner to reduce energy consumption is the need of the hour [4, 12]. The US Department of Energy listed five primary areas to focus on to develop efficient data centers; these are IT systems, environmental parameters, air management, cooling systems, and electrical systems, represented in Figure 1.3

Much advance and research is going in this area to design an efficient data center that utilizes the data center's space and increases efficiency.

1.4 BENEFITS OF ECO-FRIENDLY GREEN COMPUTING

- Green computing has resulted in a lower percentage of carbon emission to a greater extent.
- Since it has reduced the traveling of employees and customers it has lowered the greenhouse gas emission.
- Green computing laws and regulation has made the industry to increase the longevity of products.
- By the adoption of the Reuse, refurbish, and recycle technique has minimized the disposal of equipment.
- Also enhances the safety of employees in the industry by not using certain hazardous chemicals.
- Considering recycled products for the making of computer has lower the impact on the environment.

1.5 BARRIERS TO GREEN COMPUTING DESIGN

- Setting up of green computing comes at a high cost.
- Not feasible for everyone.
- Green computing technology might comprise the safety issues also.
- It requires high maintenance costs.
- Green IT may not be profit-effective for the small businessperson [13].

1.6 INTRODUCTION TO BLOCKCHAIN DEVELOPMENT

Blockchain is a technology that permits the transaction to be collected in blocks and recorded and arranged in a cryptographical manner which allows the ledger to be accessed by different servers. Every valid transaction is stored in a shared public ledger and also immutable that means the data stored can't be altered which provides high security to the data [14].

The blockchain is a peer-to-peer distributed network, encryption algorithm, and unison mechanism which is generally known as POW (proof of work). Out of all these the unison mechanisms is the main crux of the essence of blockchain. This mechanism defines the trust among the nodes in the entire blockchain network [3, 4].

Blockchain is one of the hot trending areas for Research and Development in technology. Many scholars are investigating new heights to be achieved with the help of blockchain and not only in research, but blockchain is also being used in industry in very large and significant numbers because of the security it provides [2].

The main and obvious question is why blockchain and not the ordinary ledger? The difference between the blockchain and ordinary ledger is the "Decentralization" and it's the POW which is accounted for the decentralization part of blockchain.

1.6.1 HISTORY OF BLOCKCHAIN

In 1982, the first blockchain idea was proposed by cryptographer David Chahum in his dissertation "Computer Systems Established, Maintained, and Trusted by Mutually Suspicious Groups" and then much research was focused on it, and in 1991 Stuart Haber and W. Scott Stornetta implemented where the timestamp of a document can't be altered [7].

In 2008, Satoshi Nakamoto introduced the conceptualization of the first blockchain. The design of the blockchain using a method to timestamp blocks without getting them signed by any trusted party, this design by Satoshi Nakamoto was introduced as a core component of bitcoin. The design was launched the following year by Nakamoto as a major component of cryptocurrency bitcoin, where it serves as the public logger of all transactions on the network [8, 15].

In August 2014, the file size of the bitcoin blockchain, which contained records of everything done on the network, reached 20 GB (Gigabytes). By January 2015, the size had grown to almost 30 GB, and from January 2016 to January 2017, the bitcoin blockchain grew from 50 GB to 100 GB in size. The size of the ledger had exceeded 200 GiB by early 2020.

The words block and chain were used separately in Satoshi Nakamoto's first paper but were eventually renamed as one word, blockchain, in 2016.

According to Accenture, the widespread use of innovative theory suggests that blockchains achieved a 13.5% acquisition rate within financial services in 2016, thus reaching the first phase for children. Industrial trading groups have joined the creation of the Global Blockchain Forum in 2016, which is an initiative of the Chamber of Digital Commerce.

In May 2018, Gartner found that only 1% of CIOs exhibited any type of blockchain acquisition within their organizations, and only 8% of CIOs were in a short period of time "planning or [looking at] active blockchain testing.". Figure 1.4 shows millstones of blockchain technology.

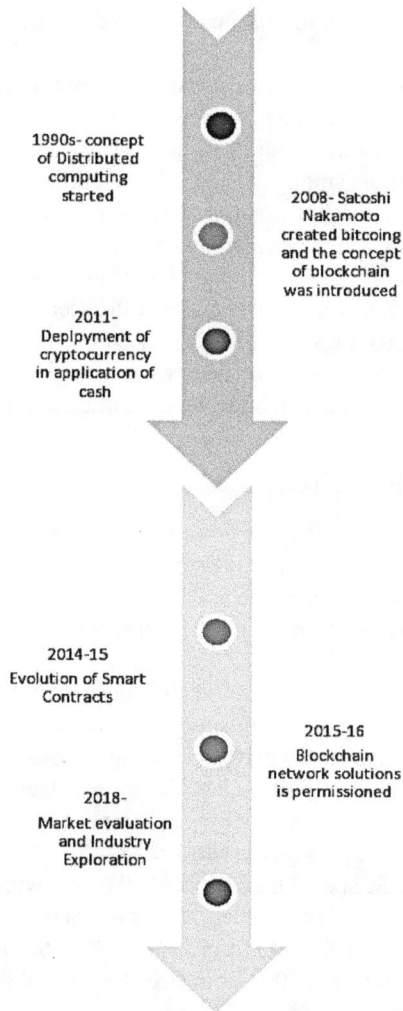

1990s- concept of Distributed computing started

2008- Satoshi Nakamoto created bitcoing and the concept of blockchain was introduced

2011- Deplpyment of cryptocurrency in application of cash

2014-15 Evolution of Smart Contracts

2015-16 Blockchain network solutions is permissioned

2018- Market evaluation and Industry Exploration

FIGURE 1.4 Milestones of Blockchain technology.

1.7 FEATURES OF BLOCKCHAIN

Blockchain is a trending topic in the technological field and day by day its functionality is touching all the areas [5, 6]. According to PricewaterhouseCoopers (PwC), blockchain technology has an opportunity to generate a value of more than $3 trillion by 2030. The main application of blockchain was into distributed ledger or cryptocurrency, but the field is now exploring due to its proof of concept; industry is now testing blockchain technology so that in the future they can fully implement their work on blockchain. So, just look over Figure 1.5, shows the various features of blockchain [16].

1.7.1 CRYPTOCURRENCY

Cryptocurrency is like virtual money or an asset like they don't exist in physical form, they are used as a medium of exchange where the transaction is stored in a ledger which gets controlled by blockchain which can review the ownership of the coin. In the starting days of blockchain or cryptocurrency, these digital currencies were not in a limelight, but now they are in so demand. To prove that we can look over the statistics, in the year 2009 the bitcoin value was around $1 but if we look now the value is massive; it is around $58,000, and according to researchers and investors, this graph of bitcoin prices is likely to grow exponentially in the near future.

1.7.2 SMART CONTRACTS

Smart contracts are the protocols that are programmed to automate the execution or control of the events that are regulated as per the terms of an agreement. The main objective of smart contracts is to create transparency and remove the involvement of the third parties this also helps to remove any fraud or any malicious exceptions. Now if we look over the working of smart contracts when the transaction is carried

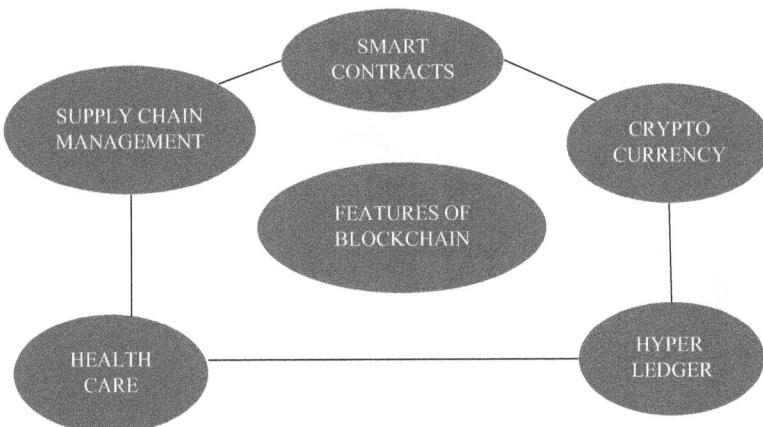

FIGURE 1.5 Feature of Blockchain technology.

out from the wallet of blockchain, the smart contract is generated, and that transaction gets stored in blockchain, then smart contract's code is executed which develop the starting of the smart contract. Once the contract is made it can't be updated. Ethereum is the most famous blockchain for running smart contracts. Ethereum smart contracts are programmed in a Turing-complete programming language called solidity. But their legality is unclear as there is very minimal usage of smart contracts but in the future, we might be able to see widespread usage of smart contracts just like how the cryptocurrency got emerged. Figure 1.6 represent the smart contract between two different parties through blockchain [17].

1.7.3 SUPPLY CHAIN

The more there is transparency in supplying the more it's beneficial for customers as well as for manufactures, blockchain has also touched the supply chain area. Blockchain purchases can help participants record price, date, location, quality, certification, and other information necessary to successfully manage a transaction. Acquisition of this information within the blockchain can increase procurement tracking, lower losses from the fraudulent and gray market, improve the visibility and compliance of export contracts, and may improve the organization's position as

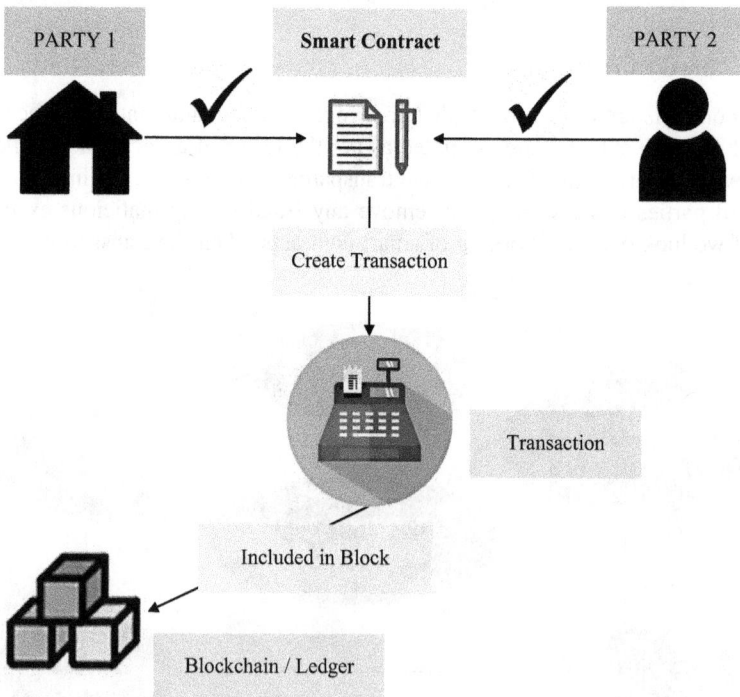

FIGURE 1.6 Blockchain-enabled Smart Contract among two parties.

a leader inappropriate production. Since there are a lot of challenges faced by supply chain management like – Lack of traceability, lack of trust, High Cost. With the help of blockchain, these constraints can be removed because you can know whom you're trading with, about the origin of raw material, about the legality of products, fair pricing [18].

1.7.4 HEALTH CARE

Wall Street Journal reported that Ernst & Young was developing a system using blockchain where they can look over all the employers and other and to keep stats of people having antibody and those who can be immune to disease. Not only that blockchain also helps in keeping the data of the patient safely, managing EMR data, personal health care management, point of care genomics management, and electronic health records data management [19].

1.7.5 HYPERLEDGER

It's like the open-source blockchains project and its tools; this was started by Linux Foundation in December 2015, and is backed up by various companies to support the development of blockchain-based distributed ledgers. The crux of the project is building a platform with the use of blockchains and distributed ledgers whose focus is on improving the efficiency of the system to those of the cryptocurrency design so that they can perform jobs on a global basis, jobs such as technical field, financial services, or supply management. In January 2018, Hyperledger released the production-ready Sawthoon1.0. In January 2019, the first long-term support version of Hyperledger fabric was announced [20].

1.8 TYPES OF BLOCKCHAIN

Figure 1.7 shows blockchain is categorized into the following types [21].

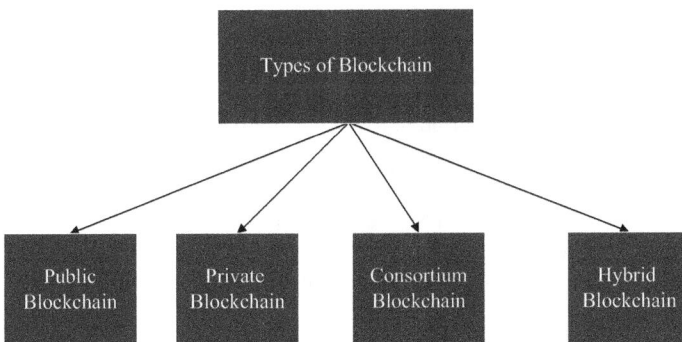

FIGURE 1.7 Different varieties of Blockchain.

1.8.1 PUBLIC BLOCKCHAIN

As the name suggests it's for everyone. In this type of blockchain, you have not required any permission to enter and do some transaction. It doesn't have any entity to look over the network, and it can't be tempered by a single person which provides more security. Cryptocurrencies like Ethereum, Bitcoin, and Litecoin use a public blockchain so that there are no restrictions to do any transaction, any person can enter into a ledger and perform a transaction in a decentralized manner [22].

1.8.2 PRIVATE BLOCKCHAIN

As its name suggests this type of blockchain can't be accessed by anybody unlike that of a public blockchain. It's also known as permission blockchain, in which there are certain entities of the organization which looks over the network and provide access to certain people and also has the power to revoke the access from anyone else. This type of blockchain is majorly used in supply chain management so that nobody apart from the team knows about the goods [23].

1.8.3 HYBRID BLOCKCHAIN

Hybrid blockchain is another type of blockchain. This is the blockchain which is included in both the public and private blockchain. The closed habitat is created instead of making it public. It's also more secure to 51% attacks and the regulations can be altered according to circumstances. In an industry related to aviation, supply chain management uses Hybrid Blockchain because it provides them trust, transparency between the user and employees [24].

1.8.4 CONSORTIUM BLOCKCHAIN

It's the semi-decentralized network with having more than one entity; they are regulated by some team or set of instructions rather than by a centralized or decentralized node. The main objective that it provides is more security, scalability, have more efficiency and regulations can be changed [25].

1.9 CHALLENGES OF BLOCKCHAIN TECHNOLOGY

Now, many companies have started to use blockchain experimentally. If we look over Deloitte's report on blockchain adoption, 53% of 1,386 executives interviewed said the use of blockchain is critical. The growth and opportunity to grow in the blockchain are tremendous, but it still has many loopholes that should be restored to attain the best use of blockchain [26].

The challenges faced in blockchain technology are shown in Figure 1.8.

1.9.1 SCALABILITY

Scalability is one of the major challenges that is faced as blockchain is getting advanced and more of it is gaining popularity, particularly in public blockchain.

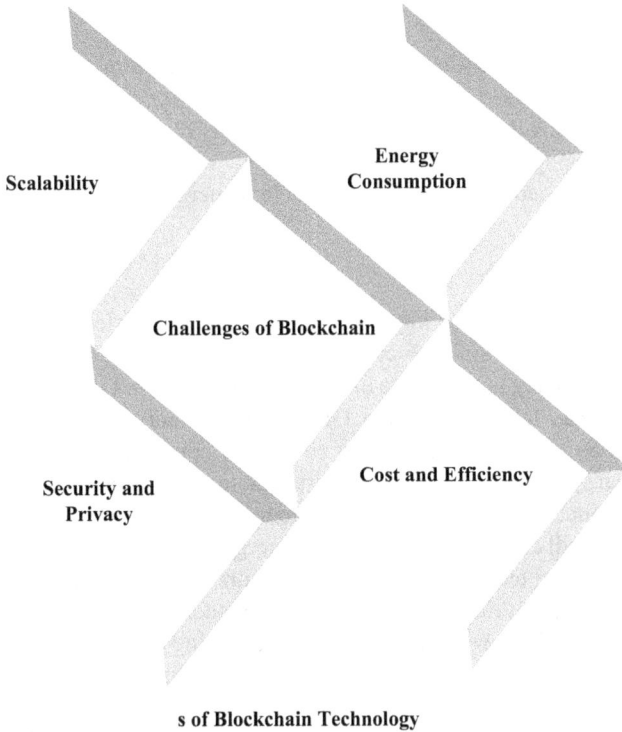

s of Blockchain Technology

FIGURE 1.8   Challenges of Blockchain technology.

The data in blockchain in bitcoin is increasing at 1 MB per block for every 10 minutes, and for Ethereum takes up to three terabytes of data. According to the researcher, by 2025 the data will be around 175 zettabytes per year [27].

1.9.2   COST AND EFFICIENCY

Since the main motive behind blockchain in the enterprise is removing the third person which reduces the cost. Since blockchain is in its infancy, and it's an expensive affair and peer-to-peer transaction is found to be expensive, this lowers the efficiency as all the nodes will tend to perform the same task in order to get to be the first one to find the solution.

1.9.3   SECURITY AND PRIVACY

The technology has provided comfort, but along with the comfort, the clouds of security and privacy are very significant. It's an open ledger this means everyone has access to your data. For the adoption of bitcoin on a widespread scale because if some sensitive information is stored then it would be more liability to keep the data safe. So, the researchers are going to enhance the security of data in the blockchain [28, 29].

1.9.4   ENERGY CONSUMPTION

Energy consumption is another major concern in the decentralization network. The most used consensus mechanism POW (proof of work) covers solving various complex puzzles which sum up to around 66.7 terawatts of energy utilization per hour. But research is being carried out and there is some other consensus mechanism that can overtake POW and could consume much less energy than POW.

1.10   CONCLUSION

Industries are now striving for ways to go green and for eco-friendly movements that both benefit nature and don't affect the outcome. Blockchain is now not only confined to cryptocurrency but, because of its distributed ledger, it is forwarding to green energy which was launched by the Global Manufacturing and Industrialization Summit (GIMS) as a green chain initiative; this combines both blockchain and renewable energy in a way that reduces carbonization and promotes going green.

Badr-Al-Olama Head of the GIMS organizing committee stated that "The Green Chain initiative will crowdsource renewable energy projects that will use 4IR [fourth industrial revolution] technologies to create the outcome of a greener planet for all"

Blockchain is a solution for electric cars too, each vehicle will be assigned its unique digital ID which will keep a record of charging and will automate the payment of charging and the vehicle is able to use those batteries as a source of electricity and provide power to run all the features of the vehicle. Adopting blockchain for green initiatives has provided a big break for sustainable development, as in the energy sector with the use of peer-to-peer solar trading, power grid management techniques are adopted, in which trading can be carried out without the help of any higher authority. There are companies in the energy sector like Acciona Energy that is pushing the limit of blockchain's potential, researching and developing ways in which blockchain-based set-up to ensure electricity supplied can be 100% renewable. The power of blockchain technology to promote environmental sustainability is greater and greater. However, technology is not yet mature enough and may not be considered a viable solution for the energy sector. Many companies are still developing proof of concept and conducting tests to test their capabilities. It may take time for businesses to realize the true power of the blockchain and move forward with digital transformation in this rapidly changing world [30, 31].

REFERENCES

1   Arjun, K.P., N.M. Sreenarayanan, K. Sampath Kumar and R. Viswanathan, "Distributed Computing and/or Distributed Database Systems," *Blockchain Platforms and Applications*, ISBN 9780367533403, Auerbach Publications, CRC Press, September 2020.

2   Dr. Saini, K., "A Future's Dominant Technology Blockchain: Digital Transformation", *in IEEE International Conference on Computing, Power and Communication Technologies 2018 (GUCON 2018)*, organized by Galgotias University, Greater Noida, September 2018, pp. 28–29. doi:10.1109/GUCON.2018.8675075.

3 Dutta, S. and S. Kavita, "Securing Data: A Study on Different Transform Domain Techniques", *WSEAS Transactions on Systems and Control*, Volume 16, 2021. E-ISSN: 2224-2856. doi:10.37394/23203.2021.16.8.

4 Dutta, S. and K. Saini, "Statistical Assessment of Hybrid Blockchain for SME Sector", *WSEAS Transactions on Systems and Control*, Volume 16, 2021. E-ISSN: 2224-2856. doi:10.37394/23203.2021.16.6.

5 Narayanan, N., K.P. Arjun and K. Saini, "A Blockchain Technology for Asset Management in Multinational Operation", *Essential Enterprise Blockchain Technology and Applications*, p. 26, CRC Press Taylor & Francis, 2021.

6 Raj, P., K. Saini and C. Surianarayanan, *Blockchain Technology and Applications* (1st ed.). CRC Press. doi:10.1201/9781003081487, ISBN-10: 0367533405, ISBN-13: 978-0367533403. 2020.

7 Saini, K. (Ed.), P.R. Chelliah (Ed.) and D.K. Saini (Ed.), *Essential Enterprise Blockchain Concepts and Applications*. Auerbach Publications. ISBN 9780367564889, 2021.

8 Dutta, S. and K. Saini, "Blockchain and Social Media", *Blockchain Technology and Applications*. Auerbach Publications, 101–114, 2020.

9 Saini, K., Chapter Title "Blockchain Foundation", *Essential Enterprise Blockchain Technology and Applications*, p. 26, CRC Press Taylor & Francis, 2021.

10 Arjun, K.P., et al., "The Advent of ML and Blockchain in Health Care", *Blockchain and Machine Learning for e-Healthcare Systems*, ISBN: 9781839531149, IET, January 2021.

11 Vikram, S., "Green Computing", *2015 International Conference on Green Computing and Internet of Things (ICGCIoT)*, Greater Noida, India, 2015, pp. 767–772, doi:10.1109/ICGCIoT.2015.7380566.

12 Kumar, N., R. Chaudhry, O. Kaiwartya, N. Kumar and S.H. Ahmed, "Green Computing in Software Defined Social Internet of Vehicles", in *IEEE Transactions on Intelligent Transportation Systems*, doi:10.1109/TITS.2020.3028695.

13 Shree, T., R. Kumar and N. Kumar, "Green Computing in Cloud Computing", *2020 2nd International Conference on Advances in Computing, Communication Control and Networking (ICACCCN)*, Greater Noida, India, 2020, pp. 903–905, doi:10.1109/ICACCCN51052.2020.9362822.

14 Butt, S., M. Ahmadi and M. Razavi, "Green Computing: Sustainable Design and Technologies", *2020 5th International Conference on Innovative Technologies in Intelligent Systems and Industrial Applications (CITISIA)*, Sydney, Australia, 2020, pp. 1–7, doi:10.1109/CITISIA50690.2020.9371781.

15 Sharma, H. and J.B. Singh, "The Effectual Real-Time Processing Using Green Cloud Computing: A Brief Review", *2020 9th International Conference System Modeling and Advancement in Research Trends (SMART)*, Moradabad, India, 2020, pp. 275–280, doi:10.1109/SMART50582.2020.9337066.

16 Guelzim, T., *Handbook of Green Information and Communication Systems || Green Computing and Communication Architecture*. pp. 209–227. 2013.

17 Sr. Jainy Jacob, M., "Relevance and Impact of Green Computing in IT", *International Journal of Engineering Research & Technology (IJERT) RTPPTDM*, Volume 3 – Issue 30, 2015.

18 Jena, Om Prakash, et al., *Green Engineering and Technology Innovations, Design, and Architectural Implementation*. CRC Press, June 2021.

19 Sadiku, Matthew N.O., *Emerging Green Technologies*. CRC Press, March 2020.

20 Al Sadoon, G.M.W., H.A. Makki and A.R. Saleh, "Green Computing System, Health and Secure Environment Management System", *2017 4th IEEE International Conference on Engineering Technologies and Applied Sciences (ICETAS)*, Salmabad, Bahrain, 2017, pp. 1–6, doi:10.1109/ICETAS.2017.8277847.

21 Khan, M.A. and Salah, K. "IoT Security: Review, Blockchain Solutions, and Open Challenges", *Future Generation Computer Systems*, Volume 82, 2018, pp. 395–411.

22 Hussien, H.M., Yasin, S.M. and Udzir, S.N.I., et al., "A Systematic Review for Enabling of Develop a Blockchain Technology in Healthcare Application: Taxonomy, Substantially Analysis, Motivations, Challenges, Recommendations and Future Direction", *Journal of Medical Systems*, Volume 43, 2019, p. 320.

23 McGhin, T., K.-K. Raymond Choo, C.Z. Liu and D. He, "Blockchain in Healthcare Applications: Research Challenges and Opportunities", *Journal of Network and Computer Applications*, Volume 135, 2019, pp. 62–75.

24 Nagasubramanian, G., R.K. Sakthivel, R. Patan, A.H. Gandomi, M. Sankayya and B. Balusamy, "Securing e-health Records Using Keyless Signature Infrastructure Blockchain Technology in the Cloud", *Neural Computing and Applications*, Volume 32, (2018), pp. 1–9.

25 Li, X., P. Jiang, T. Chen, X. Luo and Q. Wen, "A Survey on the Security of Blockchain Systems", *Future Generation Computer Systems*, Volume 107, 2017, pp. 841–853.

26 Tasatanattakool, P. and C. Techapanupreeda, "Blockchain: Challenges and Applications", *2018 International Conference on Information Networking (ICOIN)*, Chiang Mai, Thailand, 2018, pp. 473–475, doi:10.1109/ICOIN.2018.8343163.

27 Zheng, Z., S. Xie, H.N. Dai, X. Chen and H. Wang, "Blockchain Challenges and Opportunities: A Survey", *International Journal of Web and Grid Services*, Volume 14 – Issue Issue 4, 2018, p. 352.

28 Zheng, Z., S. Xie, H. Dai, X. Chen and H. Wang, "An Overview of Blockchain Technology: Architecture, Consensus, and Future Trends", *2017 IEEE International Congress on Big Data (BigData Congress)*, Honolulu, HI, USA, 2017, pp. 557–564, doi:10.1109/BigDataCongress.2017.85.

29 Chen, G., B. Xu, M. Lu, et al., "Exploring Blockchain Technology and Its Potential Applications for Education", *Smart Learning Environmental*, Volume 5, 2018, p. 1.

30 Ahram, T., A. Sargolzaei, S. Sargolzaei, J. Daniels and B. Amaba, "Blockchain Technology Innovations", *2017 IEEE Technology & Engineering Management Conference (TEMSCON)*, Santa Clara, CA, USA, 2017, pp. 137–141, doi:10.1109/TEMSCON.2017.7998367.

31 Golosova, J. and A. Romanovs, "The Advantages and Disadvantages of the Blockchain Technology", *2018 IEEE 6th Workshop on Advances in Information, Electronic and Electrical Engineering (AIEEE)*, Vilnius, Lithuania, 2018, pp. 1–6, doi:10.1109/AIEEE.2018.8592253.

2 Green Computing and Internet of Things (IoT)

Nilanjana Pradhan, Shrddha Sagar, Kavita Saini,
Nikita Bhatia, and Mayur Mathur

CONTENTS

2.1   GREEN COMPUTING

Green computing refers to the environmentally friendly use of computers and their energy. It is also known as the research and practice of designing, manufacturing, and using computers and server subsystems to eliminate them. The best example of green computing is a renewable energy source. Renewable energy sources are those that are derived from long-term reserves that can be replenished on a human timescale, such as sunlight, wind, rain, waves, and geothermal heat. They are readily available, of high environmental quality, and produce significantly less emissions. Solar power system, wind turbine programs, and geothermal power are different types of green computing.

The Energy Star voluntary labeling program was one of the first green computing programs in the USA. The Environmental Protection Agency (EPA) created it in 1992 to encourage energy efficiency in all types of hardware. The Energy Star label began to appear more frequently, especially on notebook computers and displays. In Europe and Asia, similar projects have been implemented [1, 2].

Every human being nowadays requires a machine. A machine has simplified our lives and saved us time and effort. The use of a machine, on the other hand, increases strength intake and generates additional heat. Greater use of energy and heat technologies results in higher emissions of greenhouse gases such as carbon dioxide (CO_2), which has a variety of negative effects on our climate and herbal resources. This is because we are unaware of the negative effects of computer use on the environment [3].

DOI: 10.1201/9781003107507-2

Green computing seeks to achieve economic viability while also improving how computers are used. The introduction of environmentally friendly manufacturing processes, energy-efficient computers, and improved disposal and recycling procedures are all examples of green IT practices [4].

2.2   GREEN COMPUTING AND IOT

In the coming years, there are a few information technology buzzwords that technocrats and businesses will be unable to ignore or prevent. Mobile first, artificial intelligence, blockchain, immersive experience, micro-service architecture, Zettabyte Age, robotic process automation, 3D modification, Internet of Things (IoT), quantum computing, and so on are some of these trends [2, 5]. These things will alter the way we live today and will govern the world in the future. This thesis focuses on the Internet of Things and its implementation in green computing in the current scenario, among these trending technologies [1].

The Internet of Things (IoT) is a term that refers to smart technology that link almost all and everyone in the world. The digital world is related to washing machines, refrigerators, microwaves, clocks, water taps, homewares, cooking vessels, and pots [6]. To be environmentally safe, these IoT devices must be eco-friendly, economically viable, and energy efficient [2]. IoTGC (Internet of Things-based Green Computing) is destined for such services in order to make the environment greener and smarter. The emphasis has been on the design, development, operation, maintenance, and control of energy-efficient Internet of Things (IoT) systems that are both cost-effective and user-friendly [7, 8]. Because of the numerous challenging issues such as global climate change, energy crisis, environmental issues, and so on, the field generates genuine interest among researchers [9].

According to the International Federation of Global and Green ICT (IFGICT), the green IoT (IoTGC) or green IT requires the optimum use of the internet, computers, and their resources to maintain a healthy climate. Computing systems and peripherals must be cost-effective, energy-efficient, safe, easily disposable/recyclable, and environmentally friendly. Handheld devices and large-scale data centers are needed to fit all types of systems, from mobile devices to large-scale data centers [2, 5].

Policies and plans for the planning and deployment of IoTGC devices based on real-time data will provide information about their potential environmental effects. Furthermore, if policies are decisive at various levels, substantial energy savings can be realized. Policies on energy use, automation systems, data management, and user input, among other things, will assist both end users and industries in taking the requisite corrective actions to improve IoTGC [7, 10]. The City Explorer is an example of this. In a real-life example, it's a home automation solution with layers like data collection, data processing, and services that cut energy consumption by 20%.

2.3   IOT AND HEALTHCARE MEDICAL SCIENCES

IoT involves extended network connectivity to multiple ranges of hardware devices. These devices can easily communicate over the internet with the help of sensors, transreceivers, and microcontrollers. IoT devices are embedded with different technologies. These embedded technologies are responsible for running specific tasks in a device [1, 6]. The devices can easily be monitored as well as controlled over the internet. IoT devices have a unique identity which allows remote access irrespective of the time and location i.e., it allows remote access of devices anytime and anywhere [11]. Initially, IoT was being widely used for manufacturing purposes but now the emphasis is to implement IoT in homes, offices, and in various fields like agriculture and healthcare.

Healthcare may be defined as the well-being of humans by prevention, treatment, and diagnosis of various diseases at the right time. Not only the absence of disease but healthcare is also defined as the well-being of physical and mental health of a person. It plays an important role in Gross Domestic Product (GDP) or the economy of a country. Moreover, healthcare can be achieved by medical professionals with the use of latest IoT technologies [7, 12]. Although IoT is not the only smart approach used in healthcare in today's world. it provides smart objects which act as the ultimate building blocks in the development of medical sciences and healthcare. IoT is a next generation approach and would change the lifestyle of the upcoming generation [3]. IoT has many advantages due to which it is being widely used in healthcare. Some of its advantages are:

1. Detection of disease: IoT devices have a unique feature of sensing and detecting the physical world. So, it helps in easy detection of disease. So, there are very less chances of diseases being left undetected.
2. Reduces the cost of care: Using IoT devices reduces the costs of regular visits to the doctor. Hence, this results in affordable treatment.
3. Efficient results: Due to the characteristics of IoT, the devices can record any major change in the patient's health. So, these devices provide efficient results leading to satisfaction of the patients.
4. Effective treatment: IoT devices analyses the parameters required for patients and provides effective treatment according to the requirements of the patient.
5. Less paper documents: With the help of IoT devices all the data can be stored online. Hence, there is no need to manage paper documents in hospitals and clinics.
6. Reduces human labor: Most of the work is done by machines. Hence, there is no need for a large number of humans to perform any work.
7. Saves time: In conventional times, patients had to travel long distances to reach hospitals or clinics and stand in long queues to get their treatment done. But now these devices save a lot of time by getting their treatment done at home itself. Now, patients do not need to travel long distances or stand in long queues.
8. Saves lives: Any change in the health of a patient will trigger a notification to different parties and appropriate action could be taken. So, life will be saved.
9. Reduced Maintenance costs: Now, with this technology coming into action there is no need to spend huge amounts on maintenance of medical equipment.

Therefore, IoT will prove to be a boon in the healthcare sector. It could provide effective treatment at affordable costs, thereby increasing the lifespan of patients and reducing deaths due to chronic diseases [4]. This can lead to a great development in the field of healthcare [5].

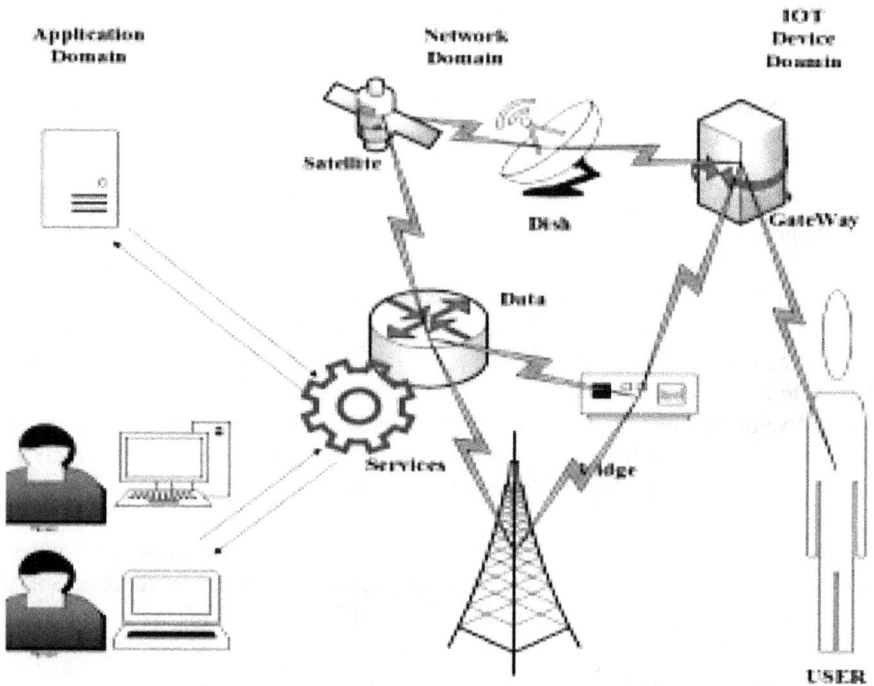

FIGURE 2.1 Extension of network devices.

Figure 2.1 [13] shows the connection and extension of various devices through the internet. In addition, it shows how the data is exchanged between devices in various domain applications [12].

There are some features of IoT which makes these devices unique and suitable in every field. Features of IoT which makes these devices suitable in every field are as follows:

(i) Self-adapting and dynamic: IoT is not only self-adapting but also dynamic in nature as IoT devices have the tendency to change themselves according to the environmental conditions they have to work in. For instance, if we take the example of closed circuit television (CCTV) cameras, they change their modes according to the intensity of light to record better pictures and videos. Even the electrocardiogram (ECG) machines used in the field of healthcare automatically adapt to changes in a patient's body temperature [5].

(ii) Self-configuring: The self-configuring feature of IoT allows IoT devices to get themselves configured on their own according to the associated devices.

(iii) Unique identity: The unique identity of these devices – i.e., internet protocol (IP) or uniform resource locator (URL) – makes these systems smart interfaces which allow communicating with users [14].

(iv) Smart interfaces: IoT devices have smart interfaces due to the employed sensing devices in them. Smart interfaces are responsible for making these devices smart and act intelligently. These interfaces help in communicating and exchanging information over the internet without human involvement [14].

(v)   Communication protocols: IoT devices have a number of protocols. Protocols are used to bolster communication between two or more devices over the internet [14].

(vi)   Interoperability: IoT devices have a unique feature of interoperability in which two or more devices can interact with each other and infrastructure without affecting the hardware and overall performance of the system [14].

(vii) Storing large amounts of data: IoT involves collecting and storing large amounts of data from various sources across different devices. The data can be structured or unstructured data. These devices help in managing large amounts of data easily and efficiently.

(viii) Analyzing data: IoT analyzes the stored data to provide information useful to the users. It helps in providing efficient and accurate results. Therefore, increasing the satisfaction rate of patients.

(ix)   Different forms of data: IoT devices can handle various types of data like audio, video, images and mathematical readings [2, 4].

With these features IoT has proved to have its implementation in several domains. Some of them are smart homes, agriculture, health and medical sciences, smart cities, industries, environment and many more [13]. Talking about the applications of IoT in the field of healthcare, the use of modern techniques has made the detection and treatment of diseases fast and efficient [7, 14]. IoT has its applications from the detection of diseases to their treatment. Moreover, it also has its application in healthcare starting from human beings to animals and all the other living beings on this earth. In healthcare IoT can be implemented in various domains:

(i)   Remote Health Monitoring – remote health monitoring means monitoring the activities of a patient when he or she is far from a doctor. IoT-based devices can be used to send the information like blood pressure, heart rate, blood group, platelet count, etc. to the doctor sitting far away. All these can be detected by the use of some special sensors and software which can be further used to store the data collected and then send it later. Due to the use of these devices the patients who are located in a place where medical help could not be sent immediately can be treated easily by using the information sent by these devices. These devices are of great help in areas like border and the areas which are backward. These devices are also used when an expert senior doctor is not physically present at the spot where the treatment of the patient is going on but instead he or she is connected with the help of video conferencing with the junior doctors present on the spot who can be supervised by the senior doctor to make the treatment successful [6].

(ii)   Real-Time Location – real-time location of the patient can be tracked with the use of different sensors, such as GPS systems. This makes it easy for the doctors as well as the ambulance to reach the patient. Moreover, if each device is installed with this system, we can easily keep track of all the devices and the patients and their real-time location. Furthermore, this will help us to provide the information about the nearest available treatment center and the location of the nearest help a patient could be provided easily.

(iii)   Fitness Programs – It is being used in many fitness programs.

(iv)   Remote Treatment of Chronic Diseases: IoT has its applications in the treatment of diseases as well. IoT simplifies the treatment of diseases and also helps in the reduction of errors in the treatment. Treatment of many diseases can be done just by using certain devices. Diseases like color blindness, cancer, kidney stones, and many more chronic diseases can be treated easily with the help of such devices.

(v)   Tracking Body Fitness – It helps in tracking the fitness of a body. E.g., Fitbit band

(vi)   Detection of Chronic Diseases – It helps in the early detection of various chronic diseases like cancers.

(vii)   Remote Medication – This domain provides medication to the patients at affordable costs and helps in the treatment of chronic diseases.

(viii)   Mobile Health: In this domain we will talk about the apps present in our mobiles. Like for tracking the intake of calories.

(ix)   Smart Hospitals: Poor infrastructure was the major problem in hospitals. With the evolution of technology, these hospitals have become smart hospitals thereby converting hospital equipment into smart beds.

These are some of the use cases of IoT in the fields of healthcare and medical sciences. In this paper we will discuss one of the domains – i.e., remote treatment of chronic diseases. We took tuberculosis (TB) into consideration which is one of the most widespread diseases in India. This disease not only attacks lungs but can also affect other parts of the body. It can spread through air when people spit, cough and sneeze [1, 2].

2.4   GREEN IOT FOR HEALTHCARE

According to a report, the World Health Organization (WHO) measures that even in developed countries like India, less than half of patients who suffer from chronic diseases take medicine as prescribed by doctors. One such disease is tuberculosis (TB). The negligence to medication has led to around 450,000 drug resistant cases in 2013 [15]. To overcome this major problem Andrews Cross and William Thies of Microsoft Research India proposed a model in 2013 named "99DOTS." The main aim of the 99DOTS model was to ensure that the patients take their medication regularly without missing any doses [15]. The project 99DOTS has been started by the Revised National Tuberculosis Control Program (RNTCP), especially for human immunodeficiency virus (HIV)-associated TB patients. Earlier, the patients had to visit their nearest direct observation therapy (DOT) centers daily in order to take their medication which they had to take under the direct supervision of the DOT providers. Direct observation therapy short-course (DOTS) comes under RNTCP which was evolved by the Government of India for the control of tuberculosis [14–16]. The main aim of RNTCP was to reduce the number of deaths that occur due to TB and to increase the cure rate of TB patients [16].

Tuberculosis in India

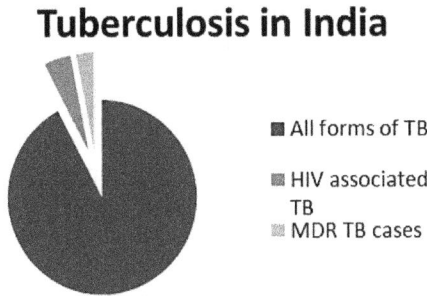

- All forms of TB
- HIV associated TB
- MDR TB cases

FIGURE 2.2   Estimated incidence of TB in India.

The patients especially the HIV associated TB patients because of their weak immunity, cannot travel long distances for taking their medication so this initiative of 99DOTS was taken for them. The initiative ensured that the patients would be able to take their medication sitting at their homes, without having to travel long distances [14]. This initiative of 99DOTS solved many problems of the patients. It was easy to implement, convenient, and affordable to use.

Figure 2.2 illustrates that out of the total population suffering from TB, 5% of the patients suffer from HIV associated TB and approx. 3% are Multi Drug Resistant (MDR) cases [3]. HIV associated TB occurs to patients who are already suffering with HIV disease. MDR patients are those patients who do not take medicine as prescribed by the doctors. Hence, their body becomes resistant to drugs [4].

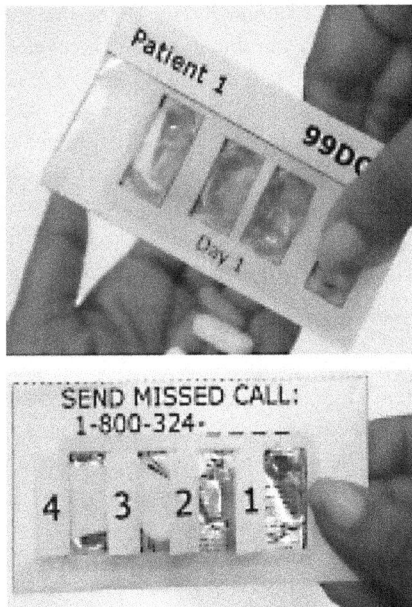

FIGURE 2.3   99DOTS model [5].

In this project patients have to call on a number each time they take their medication so that providers can monitor and maintain records. The medicines given to the patients are packed in envelopes having instructions about the dosage and a series of numbers hidden behind the pills. The hidden number which is completely unpredictable to the patient is revealed, each time a patient takes a dose of medication [12, 15]. The number revealed completes digits of a phone number, where the first few digits of the number is printed on the envelope while the other digits are the ones revealed. The patient then had to make a call to the completed phone number for the confirmation of the medicines taken which was notified to the doctor. Automated calls and reminders are sent to the patients to make sure that they do not miss their doses [14, 17].

The 99DOTS model was an advancement of DOTS therapy to get rid of the challenges faced by patients during their DOTS therapy. Further, by reading various research papers and visiting the official website of 99DOTS (https://www.99dots. org/) we have come to know that this technique has a high success rate of about 99%. This is the reason why it is termed as 99DOTS [4, 18].

2.5   CHALLENGES OF USING IOT IN HEALTHCARE

As we know, IoT devices are connected to each other on a network with the help of the internet without any human involvement [5]. These devices are also responsible for communicating and interacting between the devices and the users. Though there are distinct features and benefits that IoT devices possess, there are also many challenges in these devices. Various challenges that occur while the implementation of IoT devices are as follows:

(i)   **Networking issues**: Loss of connectivity to the internet and congestion in wireless networks. It may occur due to physical and electronic interference.

(ii)   **Complex software**: The software used in these devices must be updated from time to time, but the software used for such devices is very complex due to the complex design of these devices. So they are very difficult to modify and update.

(iii) **High installation cost**: IoT devices usually cost very high due to complex hardware and sensors used in such devices so it may sound unaffordable to some persons.

(iv)   **Security issues**: These devices collect and store data from various sources. The stored data may contain some confidential information. Cyber-attacks may be faced by such devices which can lead to unauthorized access of sensitive data.

(v)   **Lack of education**: Many people who are not educated may fail to understand the implementation and design of these devices.

(vi)   **Societal issues**: It is very important to have devices according to the requirements of the society. Their requirements change with time and so is the need to change the features of devices. Reintegrating new features will be very costly and will require high investments.

(vii) **Slow adoption of technology**: Many people are not well aware of this new technology IoT. This would lead to the slow adoption of this technology [3].

(viii) **Limited battery power**: IoT devices have a limited battery power so they are prone to failures.

(ix) **Limited storage**: Such devices usually have limited storage capacities [7]. So, a large amount of data may fail to save.

(x) **Lack of skills**: Certain skills are required in designing and implementation of these devices. Lack of skills may lead to damage to security systems [4].

All these challenges are recorded and listed from various renowned journals. These challenges can affect the overall performance of the system. As we recorded, some of the challenges are lack of education and knowledge. So, this can be solved by creating awareness among the people in different ways. Further, the remaining challenges and issues can be solved by applying various advanced standards and protocols.

2.6   IDENTIFYING THE PROBLEM AND ITS SOLUTION

In the 99DOTS project, the patients were given medicines in special medicine leaves which had a phone number written on the medicine leaf that had a missing digit. This missing digit revealed only if the pill was dispensed from the leaf. Furthermore, the person had to call on the phone number each time the medicine leaf revealed a digit for the assurance of the pill being taken by the patient [7]. But it does not give any assurance whether the pill was taken by the patient or not. What if the patient dispenses the pill and calls on the phone number revealed and throws off the pill? This was the problem which was faced by the doctors and the patients as well because it led to incorrect reports of the patients.

This problem can be solved by a neck band which can sense the shape and the weight of the pill. If it is swallowed by the patient, it can notify it to the nearest health center to make sure that the patient has taken the pill [9].

The neck bands can be given to the patients along with the medicines and they could be asked to wear it around their neck while taking the medicine [19].

These neck bands will be solar powered and will contain a micro X-ray emitter and a digital chromatographic film on the other side, so as to identify the shape and size of the medicine taken. The image captured by the film would be compared with the shape and size of the original medicine as already stored in the memory of the IoT device. Global Positioning System (GPS) technology connected to the internet can be used to send the confirmation to the doctor confirming the intake of medicine by the patient.

The proposed methodology and implementation of the neck bands is shown in the form of flowchart below. The flowchart discusses the working of the neck band and how patient's information will be stored in the database.

From Figure 2.4, we could clearly identify that according to the proposed methodology, Doctor gives the medicine to the patient. The patient is then registered against a code for tracking purposes. After the patient is registered the assistant doctor is notified about a new patient and all the details of the patient are stored in the database [1, 2]. If the patient intakes the medicine the smart band confirms the intake of medicine and sends confirmation to the assistant doctor. The assistant doctor is then notified, and the details of intake are stored in the database [17]. If the patient has not taken the medicine the smart band sends a notification to the patient after every 2 minutes. The assistant doctor provides regular reports of the patients to the

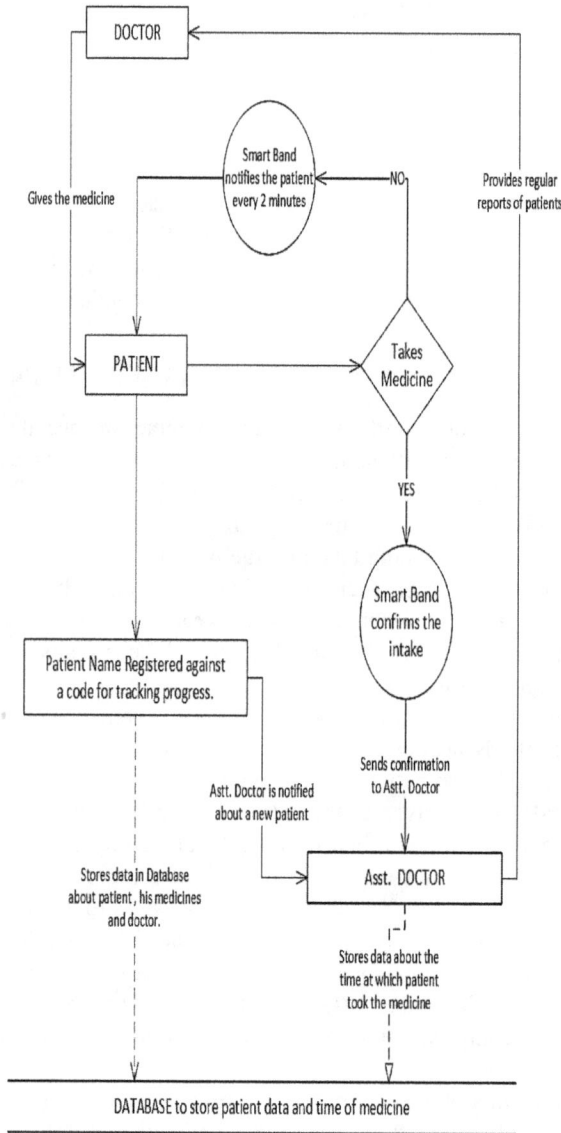

FIGURE 2.4 Proposed methodology.

doctor. All the information is stored in the database. Hence, there is no requirement of managing patient records physically. All the data will be stored online [18].

The use of neck band will not only confirm the medicine being taken but also confirm the true time of the medicine being taken by the patient. Whereas, in earlier cases we just knew the time at which the patient dispensed the medicine from the envelope. The proposed methodology tends to solve many challenges faced by TB patients during DOTS therapy.

Some of the challenges were:

- Poor health care infrastructure
- Not able to afford the services [14]
- Difficult to implement [14]
- Inconvenient for the patients to travel long distances [19].

With the proposed methodology, patients can now take their medication sitting at homes without travelling to long distances. It is easy and affordable to use. Also, the doctors can track the condition of the patient. Therefore, reducing the death rates occurring due to TB.

2.7   USE OF IOT DEVICES IN INDIA

IoT devices are widely used in India for the purpose of healthcare. IoT has become a growing trend in the field of medical sciences and healthcare in India. Nowadays, most of the clinics and hospitals in India are installed with such devices to facilitate the patients as well as the doctors.

For example, techniques such as X-ray, CT scans and MRI scans are commonly used in hospitals and clinics to treat and detect various diseases.

(i)   **X-ray**: X-ray is a technique used to detect faults in the bones of human beings. X-ray is a form of electromagnetic radiation with different wavelengths. These radiations are used in medical imaging of the bones.

(ii)   **CT scan**: CT scan uses X-ray imaging from different angles to produce a 3D image from all the 2D images. CT scan machines are installed with X-ray tubes which can be rotated and are used to take images of the organ from different angles and different positions

(iii)   **Magnetic Resonance Imaging (MRI)**: MRI uses a strong magnetic field, electric field gradients, and radio waves to generate images of the organs of our body.

IoT has increased the affordability, accessibility, and quality of care in India. These devices have led India toward a smarter way to achieve healthcare.

2.8   IOT BASED APPLICATIONS IN RURAL AREAS

The use of IoT technology is rapidly increasing in India. Due to the easy reach the urban areas are more influenced by IoT based devices for access to various services and products whereas on the other hand rural areas are not much influenced by these modern devices and their products and services [2, 7]. Therefore, IoT acts as a way between the urban and rural areas of India. IoT can play a crucial role in the development of rural areas by providing them the services same as that of urban areas in India.

By introducing these devices in rural areas, the people were benefited in the following ways:

- Easy reach to doctors
- Effective treatment
- Early detection of diseases
- People switching to modern techniques
- Increased quality of life

2.9 CONCLUSION AND FUTURE SCOPE

In this paper, we saw the various benefits and features of IoT due to which it is being widely used in many fields, especially healthcare. We have also recorded the challenges which can occur while implementing IoT. We took TB into consideration which is a very chronic disease and can even lead to death of a person.

In addition, by going through various research papers we found that only few TB patients take their medicines as prescribed by the doctors. And it was seen that there was an increase in the number of HIV associated TB patients in 2013 despite being DOTS into action, in which the patients had to visit their nearest health center to take medicine under the supervision of DOT provider. So, an initiative was taken for HIV-associated TB patients in 2013 named as "99DOTS." In this technique medication for one month is given to the patients and each time they have to call a number when they dispense the pill. Also, we came to know that the model has a high success rate of about 99%. But we identified that this model does not give any assurance whether the pill was taken by the patient or not. Patients can even throw the pill after dispensing it off. For this problem, we found that a neck band can be made which will sense the shape and weight of the medicine. A notification would be sent to the nearest health center if it is swallowed by the patient. Further, we saw how a neck band can solve all the major issues faced by TB patients and will result in the increase of curing rate of TB patients.

The advancement in IoT will detect various factors and would help in improving the quality of life. Also, this technology would help doctors to make correct decisions by which the lifespan will be boosted. Further many researchers have estimated that IoT in the field of healthcare will rise exponentially in the next 5 years. Moreover, each organization would be using IoT devices to treat their patients and maintain information about the patients as well as the diseases. Also, many research scholars have estimated that IoT will develop healthcare to an extent that there would be no disease that cannot be cured. Moreover, it has been estimated that by the end of this year almost all the hospitals and clinics would be using IoT devices.

REFERENCES

1 Zhu, C, Leung, VC, Shu, L, and Ngai, EC (2015) Green internet of things for the smart world. *IEEE Access* 3:2151–2162.

2 Saini, K, Chelliah, PR, and Saini, DK (Eds.) (2021) *Essential Enterprise Blockchain Concepts and Applications*. New York: Auerbach Publications. ISBN 9780367564889.

3 Dutta, S, and Saini, K, (2021) Statistical assessment of Hybrid Blockchain for SME sector. *WSEAS Transactions on Systems and Control* 16. E-ISSN: 2224-2856. doi:10.37394/23203.2021.16.6.

4 Narayanan, N, Arjun, KP, and Saini, K (2021) A Blockchain technology for asset management in multinational operation. In Saini, K, Chelliah, PR, and Saini, DK (Eds.) *Essential Enterprise Blockchain Technology and Applications*, pp. 152–178. New York: Auerbach Publications.

5 Saini, K (2021) Chapter Title Blockchain Foundation. In Saini, K, Chelliah, PR, and Saini, DK (Eds.) *Essential Enterprise Blockchain Technology and Applications*, pp. 1–14. New York: Auerbach Publications.

6 Dr. Saini, K. *A future's dominant technology Blockchain: Digital transformation. IEEE International Conference on Computing, Power and Communication Technologies 2018 (GUCON 2018)* organized by Galgotias University, Greater Noida, 28–29 September, 2018. doi:10.1109/GUCON.2018.8675075.

7 Raj, P, Saini, K, and Surianarayanan, C (2020). *Blockchain Technology and Applications* (1st ed.). CRC Press. https://doi.org/10.1201/9781003081487, ISBN-10: 0367533405, ISBN-13: 978-0367533403.

8 Sarhan, QI (2018) Internet of Things: A survey of challenges and issues. *International Journal of Internet of Things and Cyber-Assurance* 1(1):40–75.

9 Visalakshi, P, Paul, S, and Mandal, M (2013) *Green computing. International Journal of Modern Engineering Resolution*. In: *Proceedings of the national conference on architecture, software systems and green computing (NCASG)*, pp. 63–69.

10 Al-Sharekh, SI, and Al-Shqeerat, KHA (February 2019) Security challenges and limitations in IoT environment. *International Journal of Computer Science and Network Security* 19(2):193–200.

11 Nandyala, CS, and Kim, HK (2016) Green IoT agriculture and healthcare application (GAHA). *International Journal Smart Home* 10(4):289–300.

12 Dutta, S, and Saini, K (2020) "Blockchain and social media." *Blockchain technology and applications*, pp. 101–114. New York: Auerbach Publications.

13 Karthik, BN, Durga Parameswari, L, Harshini, R, and Akshaya, A (2018) A survey on IoT and Arduino based Patient Health Monitoring System. *International Journal of Scientific Research in Computer Science, Engineering and Information Technology* 3(1): 1414–1417. ISSN: 2456-3307.

14 Kodali, RK, Swamy, G, and Lakshmi, B (2015) *An Implementation of IoT for Healthcare*, IEEE

15 Islam, SMR, Kwak, D, Humaun Kabir, MD, Hossain, M, and Kwak, KS (2015) The Internet of Things for Healthcare: A comprehensive survey. *IEEE Access* 3:678–708.

16 Joyia, GJ, Liaqat, RM, Farooq, A, Rahman, S (2017) Internet of medical things: Applications, benefits, benefits & future challenges in healthcare domain. *Journal of Communication*, 12(4):240–247.

17 Cross, A, Rodrigues, R, D'Souza, G, and Theis, W (2014) *99DOTS: Using Mobile Phones to Monitor Adherence to Tuberculosis Medications*. Washington: Global mHealth Forum.

18 Rai, N, Singh, SP, Kushwah, SS, and Dubey, D (2017) A cross sectional study on evaluation of satisfaction level of TB Patients. *International Journal of Community Medicine and Public Health* 4(1):5–8.

19 Hemlata, KN, and Saini, SK (2011) A study on implementation of RNTCP for the community of DaduMajra Colony Chandigarh. *Nursing and Midwifery Research Journal* 7(1).

20 Oberoi, S, Gupta, VK, Chaudhary, N, and Singh, A (2016) 99DOTS. *International Journal of Contemporary Medical Research* 3(9):2760–2762.

21 The 99 DOTS Website. Available at: https://www.99dots.org/ Accessed: 18 January 2019.

22 Jindal, F, Jamar, R, and Churi, P (April 2018) Future and challenges of Internet of Things. *International Journal of Computer Science and Information Technology* 10(2):13–25.

3 Coalescence of Neural Networks and Blockchain

Mohit Dayal, Ameya Chawla, and Manju Khari

CONTENTS

3.1 INTRODUCTION TO AN ARTIFICIAL NEURAL NETWORK

Neural networks or artificial neural networks, as the name suggests, is a connection of nodes which we call neurons or artificial neurons. This structure is comprised of many artificial neuron layers connected together which is inspired by the structure of the human brain. Neural networks are considered better than many other machine learning algorithms due to their capability of finding complex relations between the input data and the output data.

DOI: 10.1201/9781003107507-3

Neural networks are part of deep learning which is a subset of machine learning where we try to replicate the structure of the human brain. Neural networks are one of the most used machine learning models due to their high computational power and wide range of applications like CNN – convolutional neural network which is used in deep learning problems related to images, RNN – recurrent neural network used in deep learning problems related to speech, etc.

Blockchain is an expanding record list which are termed as blocks and linking of these blocks are done using cryptography which form chain-like structures which give it its name blockchain. Each block which is added contains following information like hash code, a timestamp, and transaction data all these are stored in the block and hash code is a cryptographic code which is of the last block which is added to the chain. This type of structure can not be altered as the block is added for each transaction and it is design in the way to be effective against modification. The most common use of blockchain technology is cryptocurrency like bitcoin.

Blockchain enables data systems which can be either used in healthcare or student education sector all can be implemented using blockcahin technology which makes easier to add user data and is more secure. Then neural network can be trained on that type of data to make more easier method to analyze the data fit model on it rather than generic way to collect data [1].

Objective: This chapter will give a brief overview of neural networks to the reader with full understanding about the terms related to the neural network, architecture of the neural network and derivation of equations related to neural network. There will be a brief overview of blockchain technology and real-life application of blockchain technology and neural networks.

3.2   ARCHITECTURE OF AN ARTIFICIAL NEURAL NETWORK

Neural network architecture is mainly divided into basic components of layers having neurons and neurons having weights, biases, and activation functions.

3.2.1   LAYERS

Artificial neural networks have basically three types of layers:

3.2.1.1   Input Layer

First layer of a neural network which takes input from the data and where each neuron gets input from the feature of the given data and input layer neurons are equal to features in the given data.

3.2.1.2   Output Layer

Last layer of a neural network which gives the output of the model, neural network output can be of either regression or classification. A neural network with regression output will have only one neuron in the output layer while in classification output layer neurons depends on number of classes. In regression a single value will be returned by the neural network and in case of n classes n probabilities are given where each probability signifies the chance of that class being the predicted class. Maximum probability class is taken as the final predicted class.

3.2.1.3 Hidden Layers

All the layers which are between input layer and output layer are considered as hidden layers. Each layer is connected with its previous and next layer as it takes input from previous layer and gives output to the next layer. These layers are not visible to user as the user only gives input and expects output so the layers are hidden from perspective of the user (Figure 3.1).

Yellow-colored neurons represent the input layer neurons which takes the input, pink-colored neurons are hidden layer neurons which take input from the input layer, and blue-colored neuron is the output layer neuron which gives the predicted class or regression output.

3.2.2 NEURONS

The neuron is the smallest unit of the neural network which takes input from all previous layers of neurons, processes it, and sends it to all neurons in the next layer. Let's understand what a neuron does when it gets input from the previous layers, first it calculates the summation of product of each input value and weights assigned for that value in the particular neuron and then add bias value to it after it activation function is applied and that output is sent to next layer neurons. The activation function transforms the summation of product of weights and inputs (with addition of bias) so that it can be easily recognized by the next layers whether it is a useful input or not (Figure 3.2).

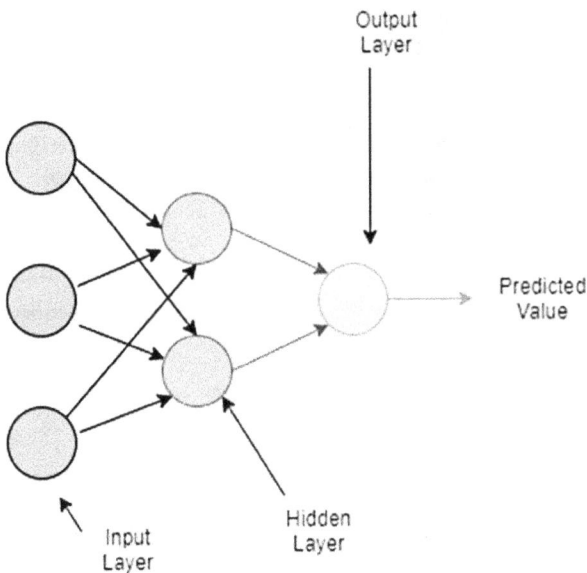

FIGURE 3.1 Neural network structure of a classifier.

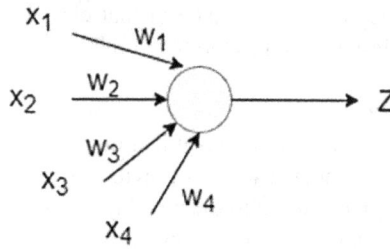

FIGURE 3.2 Detailed look at neurons in layers.

$$y = w_1 x_1 + w_2 x_2 + w_3 x_3 + \ldots + w_n x_n + b$$

$$z = \text{Activate} \left(y \right)$$

$x_1, x_2, x_3 \ldots x_n$ are the n inputs which are received by the neuron and $w_1, w_2, w_3 \ldots w_n$ are the weights assigned to the inputs received and b is the bias assigned to the neuron. The neuron transmits z to the next layer of neurons.

3.2.3   IMPORTANCE OF BIAS AND WEIGHTS IN A NEURAL NETWORK

3.2.3.1   Weights

Weights are used with each feature given as an input to the neuron and weights tell which feature is more important. As an example, if the weight of one feature is more than another then that means that feature is more important while predicting the output than the other feature with less weight value. Negative value of weights determine that the feature is inversely proportional to the target value which we are predicting.

3.2.3.2   Biases

Bias is a constant value added to the summation of product of weights and features. Bias is added to make the activation function to shift its direction in either left or right, as an example if we are making the decision that if summation value is greater than 0.5 we will assign class 1 and if less than 0.5 we will assign class 0. This threshold value 0.5 which we have set can be used in comparison and this threshold constant is used in decision making. Let's consider the case when the summation is greater than 0

$$\sum w_i x_i > 0.5$$

$$\sum w_i x_i - 0.5 > 0$$

As we can now see, it is easier for the activation function to fit properly on the given data where boundaries are defined for the summation being greater than 0 in the case of class 1 and the summation being less than 0 in class 0. Bias value can also be considered as intercepts; it doesn't change the slope but changes the position of the graph so that it can fit as per our needs.

3.2.4 ACTIVATION FUNCTIONS

Activation functions are mathematical functions which determines the output of the neuron. Different activation functions are used for different types of problems and activation function used also depends whether problem is regression problem or classification problem. Gradient of weights and biases plays a very important role while choosing the activation function, Gradient is defined as rate of change in a variable respect to other variable and gradient of error with respect to weights and biases which we achieve after training is used to update biases and weights to reduce the error in prediction [2].

3.2.4.1 Linear Function

This is the simplest activation function where a constant is multiplied without variable.

$$z = ay$$

$$z' = a$$

Gradient of z with respect y shows a constant value which means the weights and biases will get update by a constant factor always and error will remain the same.

3.2.4.2 Sigmoid Function

The sigmoid function is a mathematical function which transforms the value between range [0,1] and it is a non-linear activation function.

$$z = \frac{1}{1+e^{-y}}$$

$$z' = \frac{1}{1+e^{-y}} * \left\{ 1 - \frac{1}{1+e^{-y}} \right\}$$

Gradient of the sigmoid function shows the gradient is dependent on the variable y and as y tends to either $+\infty$ or $-\infty$ the gradient will become 0 and will be the same as the linear function gradient for higher values.

3.2.4.3 Hyperbolic Tangent Function

This is a non-linear function which is derived from the sigmoid function by making small changes in the sigmoid function.

$$z = 2 * \frac{1}{1+e^{-2y}} - 1$$

This function gives range from [−1,1] and is symmetric about origin.

$$z' = 1 - \left\{ 2 * \frac{1}{1+e^{-2y}} - 1 \right\}^2$$

As y tends to either $+\infty$ or $-\infty$ the gradient will be become 0 and will be as same as the linear function gradient for higher values.

3.2.4.4   Rectified Linear Unit

This is a non-linear function where if our value of y is greater than 0 then y is returned else 0 is returned and this property is advantage of rectified linear unit that it only send positive values or 0.

$$z = \text{maximum } (0,y)$$

$$z' = \text{maximum } (0.1)$$

Gradient in this case is always constant which means the means the weights and biases will get update by a constant factor always and error will remain same.

3.2.4.5   Exponential Function

The exponential function is a non-linear function where if y is greater than equal to zero it remains same else it is transformed using exponential function.

$$z = y \qquad \text{if} \qquad y > 0$$

$$z = a * \{e^y - 1\} \qquad \text{if} \qquad y <= 0$$

For positive values it is a linear curve and for negative values it is a exponeential curve. This function allows negative values as output which makes the model learn faster and better.

$$z' = 1 \qquad \text{if} \qquad y > 0$$

$$z' = a * \{e^y\} \qquad \text{if} \qquad y <= 0$$

Gradient is constant for values of y greater than 0 and exponential for values less than 0 which signifies that in case of negative values of y the model weights and biases will be updated.

3.3   WORKING OF AN ARTIFICIAL NEURAL NETWORK

When we train an artificial neural network the data is transmitted in each epoch twice through the model once transmitted forward and once backward. The transmission of data through the whole model is called propagation.

3.3.1   FORWARD PROPAGATION

Data is transmitted from input layer neurons to the output layer and output is predicted from the output layer. Output given from output layer marks the end of forward propagation. Each neuron performs the same function by calculating weighted

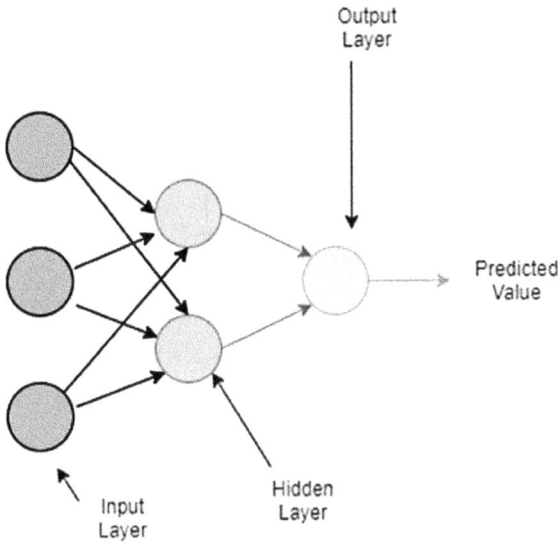

FIGURE 3.3 Forward propagation.

sum with addition of bias and then passing it to the activation function. The output is given to each neuron in the next layer [3] (Figure 3.3).

Yellow, pink, and blue arrows show how the data is transmitted in the forward propagation.

3.3.2 BACKWARD PROPAGATION

This is the reverse process of forward propagation where process goes in backward direction. Backward propagation is based on the principle of error where we first calculate error in our prediction, then we try to propagate backward and check how the error is affected by specific neurons' weights and bias by calculating the gradient of error in relation to the variable weights and bias and then update them accordingly [4].

Let's consider an example in Figure 3.4:

Blue arrows show how data is transmitted from the output layer to hidden layer with pink-colored neurons. Let's find the loss function which is defined as square of difference in actual value and predicted value.

$$E = \left\{ z_p - z_a \right\}^2$$

z_p is the value predicted by the neural network and z_a is the actual value expected at the output, where z_a is a constant and z_p is calculated using the activation formula in last neuron.

$$z_p = \text{Activation}\left(y_p \right)$$

$$y_p = \sum w_i x_i + b$$

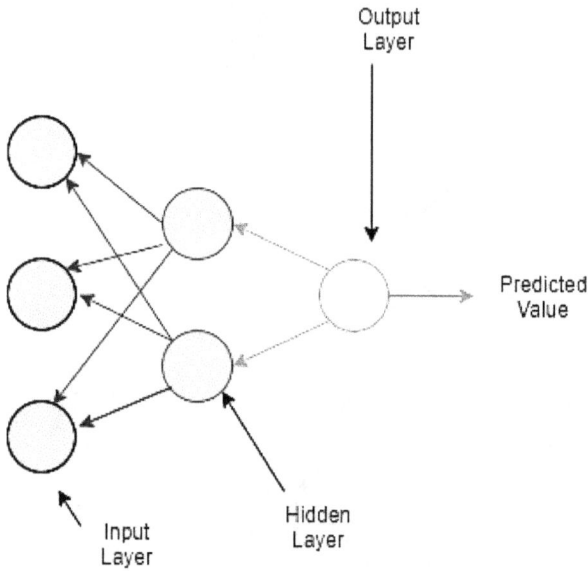

FIGURE 3.4 Backward propagation.

Now let's consider the weights in the next Figure 3.5:

Let's consider the output neuron (blue neuron): it is receiving 2 weights W_4 and W_3 and has bias value b_3 so while back propagating the values of them will be updated as:

$$W_4 = \left\{ W_4 - \eta * \text{gradient} \left(\text{loss} \right) \right\} \quad \text{gradient with respect to } W_4$$

$$W_3 = \left\{ W_3 - \eta * \text{gradient} \left(\text{loss} \right) \right\} \quad \text{gradient with respect to } W_3$$

$$b_3 = \left\{ b_3 - \eta * \text{gradient} \left(\text{loss} \right) \right\} \quad \text{gradient with respect to } b_3$$

where η is the learning factor which is a hyper parameter and the goal of the algorithm is to update weights and bias for each of the neurons and to optimize the error.

Gradient of loss with respect to W_4 is calculated as:

$$\frac{\partial E}{\partial W_4} = \frac{\partial \left\{ y_p - y_a \right\}^2}{\partial W_4}$$

$$\frac{\partial \left\{ y_p - y_a \right\}^2}{\partial W_4} = 2 * \left\{ y_p - y_a \right\} * \frac{\partial y_p}{\partial W_4}$$

Let's consider linear activation function:

$$y_p = a * \left(W_4 x_4 + W_5 x_5 + b_3 \right)$$

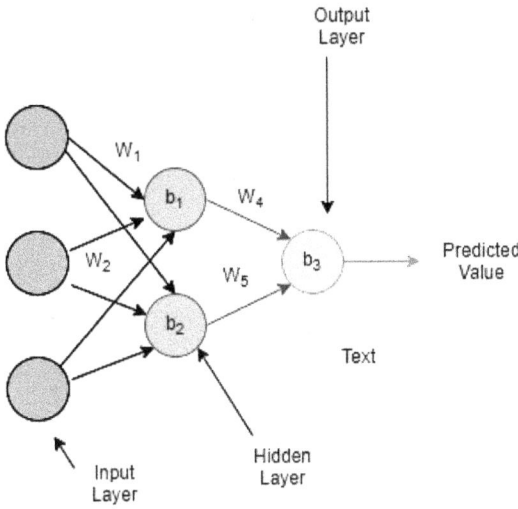

FIGURE 3.5 Backward propagation.

$$\frac{\partial y_p}{\partial W_4} = ax_4$$

W_3, x_4, x_3, b_3 are considered constants while doing partial derivative with W_4. Similarly we can calculate gradient of error due to bias.

$$\frac{\partial y_p}{\partial b_3} = a$$

Let's consider example of pink neuron with bias b_2. Let's consider it is sending z_2 as input then:

$$\frac{\partial E}{\partial b_2} = \frac{\partial E}{\partial z_2} * \frac{\partial z_2}{\partial b_2}$$

z_2 can be considered as x_5 sent as input which implies

$$\frac{\partial z_2}{\partial b_2} = a$$

$$\frac{\partial E}{\partial x_5} = \frac{\partial \{y_p - y_a\}^2}{\partial x_5}$$

$$\frac{\partial \{y_p - y_a\}^2}{\partial x_5} = 2 * \{y_p - y_a\} * \frac{\partial y_p}{\partial x_5}$$

As we have already calculated y_p then substituting it and we can calculate following gradient.

$$y_p = a * \left(W_4 x_4 + W_5 x_5 + b_3 \right)$$

$$\frac{\partial y_p}{\partial W_4} = W_5$$

using the above result

$$\frac{\partial E}{\partial b_2} = a W_4$$

Hence we can calculate gradients for all weights and biases of each and every neuron and update them in each epoch.

3.4   IMPLEMENTATION OF NEURAL NETWORK USING PYTHON

There are many ways to implement a neural network either by making our own class and defining all weights, biases, and activation functions; we will be using predefined modules in Python. We will be using Scikit-learn library to create our neural network module. Data which we will use in this implementation is Titanic dataset. Many fields have practical implementation of ANN [5, 7] (Figures 3.6–3.9).

```
#Used to build model
from sklearn. neural network import MLPClassifier
#Used to s plit given training data to t e st and
#train set
from sklearn. model selection import train test split
#Used to read the data set
import pandas as pd
#Reading training data set from f i l e saved locally
train=pd. read csv ("train.csv")


#This method prints information about a DataFrame
#including the index dtype and columns , non-null
#values and memory usage .
```

FIGURE 3.6   Initial steps of data preprocessing and importing required libraries. (Continued)

```
train.info ()
#This method generate Generate descriptive st ati sti c s .
train.describe ()


Dropping all the columns which have no relation with the
#death of the person
train=train.drop (["PassengerId","Ticket","Fare",
"Embarked","Name","Cabin"],axis =1)
#dropping all rows with null values
train=train.dropna ()
#checking info afte r dropping null value rows and columns
```

FIGURE 3.6   (Continued)

```
train.info ()
#Label encoding the categorical data
train x ["Sex"]= LabelEncoder (). fit transform ( train x ["Sex"])
#Dividing the training data into train x and train y
train x =train.drop (["Survived"],axis =1)
train y =train["Survived"]
#Further dividng them to train and t e st data set s
train xx = train x [ train x .columns ][:640]
train yy = train y [:640]
test xx = train x [ train x .columns ][640:]
test yy = train y [640:]
model= MLPClassifier (solver='sgd',
alpha =1e-5,hidden layer sizes =(2 ,1 ), random state =1,
activation ="logistic", max iter =100)
# f it t i n g the model for our training data


model.fit( train xx , train yy )
#te sti n g the model for t e st accuracy
model.score( test xx , test yy )
```

FIGURE 3.7   Initializing the model and training it and testing it.

```
<class 'pandas.core.frame.DataFrame'>
RangeIndex: 891 entries, 0 to 890
Data columns (total 6 columns):
 #   Column     Non-Null Count   Dtype
---  ------     --------------   -----
 0   Survived   891 non-null     int64
 1   Pclass     891 non-null     int64
 2   Sex        891 non-null     object
 3   Age        714 non-null     float64
 4   SibSp      891 non-null     int64
 5   Parch      891 non-null     int64
dtypes: float64(1), int64(4), object(1)
memory usage: 41.9+ KB
```

FIGURE 3.8   Using method info for information about data.

```
<class 'pandas.core.frame.DataFrame'>
Int64Index: 714 entries, 0 to 890
Data columns (total 6 columns):
 #   Column     Non-Null Count   Dtype
---  ------     --------------   -----
 0   Survived   714 non-null     int64
 1   Pclass     714 non-null     int64
 2   Sex        714 non-null     object
 3   Age        714 non-null     float64
 4   SibSp      714 non-null     int64
 5   Parch      714 non-null     int64
dtypes: float64(1), int64(4), object(1)
memory usage: 39.0+ KB
```

FIGURE 3.9   Using method described for statistics of the data.

The model trained obtained an accuracy of 60.81%. The accuracy depends on many features like choosing activation functions, number of neurons in each layer, number of layers, and optimizing algorithm, etc.

3.5   LITERATURE REVIEW

1. DeepRing: Protecting Deep Neural Network With Blockchain

Deep neural networks have vast variety of applications and blockchain technology is also accepted as one of the most secure techonolgy in cybersecurity field. This paper presents the combination of both the worlds by first making a deep neural network model and then securing the model using blockchain technology [8].

2. **Machine Learning Adoption in Blockchain-Based Smart Applications: The Challenges, and a Way Forward**
Security is one of the most important aspects of any technology; there are attacks possible on one of the most secure blockchain technologies so to protect it machine learning models are used to analyze the attacks on the blockchain and to make it more secure [9].

3. **An AI-Based Super Node Selection Algorithm in Blockchain Networks**
One of the major problems faced in blockchain technology is the large consumption of electricity. Application of artificial intelligence (AI) can help in reducing the unwanted hash operations performed and hence can help in saving electricity [10].

4. **Machine Learning In/for Blockchain: Future and Challenges**
Machine learning and blockchain are both among the most important technologies used in the world and both are data-driven technologies. The combination of machine learning and blockchain to make more efficient and effective solution for data-driven technologies [11].

5. **Secure Decentralized Peer-to-peer Training of Deep Neural Networks Based on Distributed Ledger Technology**
Privacy is a key factor while making secure systems and blockchain is one of the most secure transaction technologies. Using deep learning techniques, a secure peer-to-peer-based framework can be made on based on blockchain technology to make a more secure system for transactions [12].

3.6   OPEN CHALLENGES

There are many open challenges related to the technology of blockchain but the most common challenge is to convince the public to adopt it as most users have a misconception in their mind relating blockchain technology to cryptocurrency which most people think as an illegal currency or currency used by hackers and fraud users. Another main challenge faced in the technology is speed as this system used for transactions is very slow as compared to other transaction systems used in modern-day technologies.

REFERENCES

1 M. Dayal, N. Singh, Indian health care analysis using big data programming tool, *Procedia Computer Science* 89 (2016) 521–527.
2 M. M. Lau, K. H. Lim, *Review of adaptive activation function in deep neural network*, in: *2018 IEEE-EMBS Conference on Biomedical Engineering and Sciences (IECBES)*, IEEE, 2018, pp. 686–690.
3 G. Cerri, M. Cinalli, F. Michetti, P. Russo, Feed forward neural networks for path loss prediction in urban environment, *IEEE Transactions on Antennas and Propagation* 52 (11) (2004) 3137–3139.

4 Z. Tang, O. Ishizuka, H. Matsumoto, *Backpropagation learning in analog t-model neu-ral network hardware*, in: *Proceedings of 1993 International Conference on Neural Networks (IJCNN-93-Nagoya, Japan)*, Vol. 1, IEEE, 1993, pp. 899–902.

5 M. Khari, A. K. Garg, R. G. Crespo, E. Verdu, Gesture recognition of rgb and rgb-d static images using convolutional neural networks, *International Journal of Interactive Multimedia & Artificial Intelligence* 5 (7) (2019) 22–27.

6 M. Dua, R. Gupta, M. Khari, R. G. Crespo, Biometric iris recognition using radial basis function neural network, *Soft Computing* 23 (22) (2019) 11801–11815.

7 Y. H. Robinson, S. Vimal, M. Khari, F. C. L. Hernandez, R. G. Crespo, Tree-based con-volutional neural networks for object classification in segmented satellite images, *The International Journal of High Performance Computing Applications* (2020) 1094342020945026.

8 A. Goel, A. Agarwal, M. Vatsa, R. Singh, N. Ratha, *Deepring: Protecting deep neural network with blockchain*, in: *Proceedings of the IEEE/CVF 16 Conference on Computer Vision and Pattern Recognition Workshops*, 2019, pp. 2821–2828.

9 S. Tanwar, Q. Bhatia, P. Patel, A. Kumari, P. K. Singh, W.-C. Hong, Machine learning adoption in blockchain-based smart applications: The challenges, and a way forward, *IEEE Access* 8 (2019) 474–488.

10 J. Chen, K. Duan, R. Zhang, L. Zeng, W. Wang, An AI based super nodes selection algorithm in blockchain networks, arXiv preprint arXiv:1808.00216.

11 F. Chen, H. Wan, H. Cai, G. Cheng, Machine learning in/for blockchain: Future and challenges, arXiv preprint arXiv:1909.06189.

12 A. Fadaeddini, B. Majidi, M. Eshghi, Secure decentralized peer-to-peer training of deep neural networks based on distributed ledger technology, *The Journal of Supercomputing* 76 (2020) 1–15.

4 Blockchain in Administration
Advances and Implications of Decentralized Applications for Sharing Information

M. Kumar, Himanshu Dubey, and K. Kumar

CONTENTS

4.1 INTRODUCTION

The broad determination of blockchain technology (BCT) is observed as the main furthermost significant technology drifts that will affect commerce and culture in the coming years. BCT has materialized as a possibly disorderly, general-purpose technology for organizations and administration to support transactions and information that needs trust and authentication.

Vos et al. (2017) discussed BCT which carries the equivalent data at various nodes and the data will only be put in as per consent to have the nodes among the nodes. A fresh dealing can be appended; however, data cannot be detached. Stowing information comes via transaction in distributed nodes which is named as a distributed ledger. This lessens the need for a principal performer and the danger of operation or arrangement letdown all nodes have the complete information.

Dimitrov (2019), found that blockchain (BC) can be consumed for any alteration of tenure and the stowage of significant information and leaflets like "licenses,"

DOI: 10.1201/9781003107507-4

"certificates," "government-related decisions and legislation.". Classically, data kept in a BC are transactional data like the tenure of land admin, birth and wedding certificates, lorry registries, commercial licenses, didactic certificates, scholar loans, communal welfares and ballots. BCT has the likelihood to deliver improvements to administration and civilization and can extant the following stage in "e-government" growth, as they allow abridged charges and intricacy, pooled right-hand courses, better discoverability of audit test and confirmed reliable record card. In this paper authors are intense on level technique, talking the technological complexity of using BC technology for peer-to-peer (P2P) manners.

There is need to offer chances for designing transaction and data altercation courses in the secluded context. In juxtapose, barely any study is attentive to BC technology and its aptitude to discourse communal requirement. In this paper, authors make an impression of vital welfare and recognize fresh characters for administration to achieve BC technology and safeguard their benefits. The author's main objective is to donate to a more proved conversation about BC in administration by sketching the care to facets that are understated and need further study.

4.2 LITERATURE REVIEW

Vos et al. (2017): the authors discuss the potential usage of blockchain technology in administration, with a summary of about latest advances in the arena of blockchain technology as an instance of methodical maturity, and administration. The association between Being Individual(s), Right(s) and Entity(s) in a government system is the foundation for the description of mandatory functionality, assumed the intricacy within these three fundamentals: uniqueness of an individual, lawful variety in objects. The paper investigates how certain principles of good administration, which include transparency, accountability, security, and regulation, are possible by blockchain technology. It is concluded that the method does not to be mature adequate for operation in administration in this instant.

Paik et al. (2019): the authors set out to upsurge the consideration of blockchain technology as a data store and to endorse a systematic method of overcoming its too bulky software systems. Jha et al. (2019): in this paper, the authors classify the mutual layers of a typical software system with data stores and conceptualize each layer in blockchain terms. Another, authors inspect the location and movement of information in "blockchain-based" operation. Finally, authors scrutinize the information administration complexity in blockchain with relationship assurance in terms of privacy and security.

Grover et al. (2019) describe the blockchain and its dispersion looks to be changeable for dissimilar industries. The objective of this study is to discover the blockchain technology dispersal in variety industries through a mixture of academic literature and social media.

Farouk et al. (2020), elaborate of a combination of data, network, and blockchain technology in healthcare. The healthcare data are come to be an imperative element in this sector. There is need to secure and safe transmission of medical data among the organization. The big challenge is to collect the information about patient diseases which has become essential to restrict the pandemic in the world. There are numerous ways of doing this, although it is not possible to obtain complete and temporal information at a given time. To bring the transparency between the organizations around the world, there is need to share the data directly across the world. This can be done only via technology which includes a combination of IoT network and blockchain which are decentralized in the system and have the capability to store the data in distributed manner.

Gökalp et al. (2018): the main objective of this paper is to suggest a blockchain framework in medical which include all shareholders in healthcare system to examine prospects and contests by giving a unified blockchain design.

Bell et al. (2018) discuss numerous sectors of healthcare system and safety that could be improved using blockchain technologies. These comprise of detecting phenomenon by distributed application, medical tests, pharmaceutical detecting, and health insurance. The detectable appliance is capable to trace their purpose within the infrastructure of blockchain. The information collected can be utilized advance security and safeguard of patient by getting the depth detail of market retail analysis to enhanced efficiency effectively. In this paper, author present latest sectors of pharmaceutical detectability, data sharing, clinical tests, and device tracing.

Hoy (2017): the authors discuss blockchain technology that is a comparatively a new technology and used to prove and supply the transaction records for virtual cryptocurrencies like Bitcoin. The system used in this process is distributed and redundant in nature, which makes difficult for the transactions to be revoked or faked.

Dimitrov (2019) gives theoretical consideration to the practical basics of the budding of blockchain technology in healthcare system, which is required to comprehend precise blockchain application, estimate commercial cases including blockchain startups, or trail the argument regarding its anticipated economic effects.

Casino et al. (2019) presents a systematic literature review of the blockchain technology used in the multiple domains. This paper demonstrates the ongoing state of blockchain technology and also present the sorting of blockchain focused applications in the various sectors such as supply chain, healthcare, business, IoT, privacy, and data management. This paper also helps in the emerging areas of research in the blockchain technology.

Myeong and Jung (2019): the authors discuss the increasing interest in the blockchain technology in the administration which can increase the economic efficiency, security, and decentralization in administration. Authors also tells use this technology in the upcoming future in the public sector such as e-government which can enhance the administrative process in the government in the various countries to enhance the administrative process, security, and also in the data management in the governance sector.

4.3 LITERATURE GAP

In the literature review discussed in this paper there are the various limitations or gaps in the literature which are as follows:

1. **Scalability Problem**: In the existing research of blockchain in various management systems there is the limited number of transactions are processed in which the number of blocks used in the blockchain are limited in frequency and size.

2. **Lack of Transparency**: There is a lack of transparency or we can say lack of centralized transaction process in the existing blockchain of management system which makes the lack of trust in the management systems by the people.

3. **Absence of Trust in Network**: In the various management systems, there is a lack of trust in the network which is used in the organization, institutions, etc. which makes the management system more vulnerable to cyber-attack.

4. **Security**: To run blockchain technology in the existing management system there is the need of advanced security features such as encryption to encrypt every transaction process and hashing method to link the encrypted transaction process to the old transaction process.

5. **False Traceability**: There is the high false traceability rate in the supply chain of existing or traditional management system which can make the several problems such as theft, counterfeit, and loss of goods etc.

6. **Cost**: Organizations spend a lot of money to manage the existing management system and the organization needs to reduce the cost and use the money for development of the organization such as improvement in the existing system, improvement in infrastructure, develop new techniques, etc.

4.4 PROPOSED FRAMEWORK

Blockchain technology is the well-versed broad technology which is used to handle data and digital assets in distributed mesh networks, and the use of this technology in administration management systems can solve various problems in administration which we discussed in the previous section and enhance the administrative process in states, countries, or in the world. In the proposed framework of blockchain in administration which is given below in the Figure 4.1 we are going to solve the corruption level in the revenue department which is part of any administrative management system.

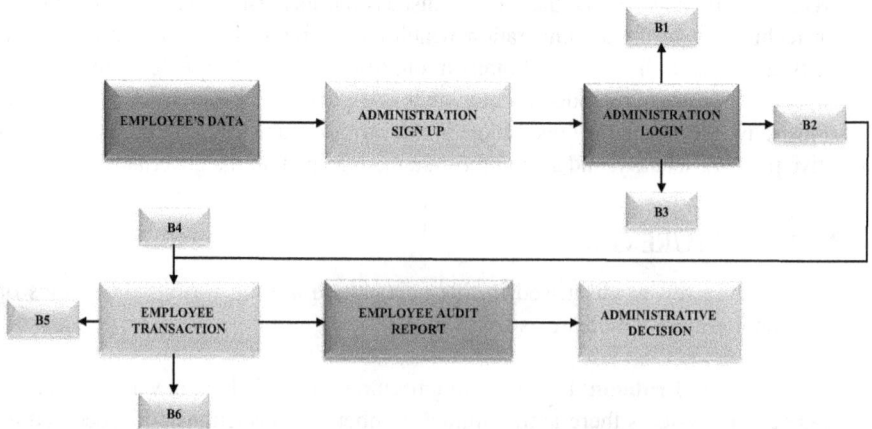

FIGURE 4.1   Proposed framework.

The above proposed framework in the Figure 4.1 illustrates the use of blockchain in the administrative process of revenue department in which we use the employee data and analyze the corruption level in the employee transaction with the help of the audit report and takes the administrative decision by the revenue administrator. This proposed framework can also be used in the various administration management system, healthcare, businesses, and in e-government etc. which makes the enhancement in the administrative process.

The proposed framework discussed about the methods of developing the administration systems is based on the machine learning approaches which include signature-based administration system and profile-based administration system. The authors analyze that the signature-based approaches are based on the current and previously stored information that uses the known pattern of data to identify the corruption in the revenue department. The drawback of the signature-based approach is that it is limited to find only known pattern-based corruption. The second is the anomaly-based approach which is used to predict the corruption based on the samples which are unknown patterns and not decided previously. For finding the unknown pattern the anomaly-based deep learning approaches are needed that is convolution neural network, recurrent neural network, and artificial neural network. There is a need for advanced research in the field of tracking systems that could be beneficial for many people and can generate employee audit report for the employee within a fixed period. Figure 4.1 shows an administration system which allows integrating data from the multiple data center in multiple formats. The user sends the request to the administrative officer with an administrative problem statement where the administration has some criteria that need to follow by an employee in the department to generate the identity with sign up and log in, and to proceed the request to an intelligent system of the administrative system for generating the report with the history of the employee transaction. The above figure represents the administration system that is connected with employees in the revenue department of administrative system through the IoT network. Employee that connects with the IoT network and communicate with the administration intelligent system by the mutual relationship between the existing employee transaction and recent problems related to administration system.

4.5   DATA FLOW ANALYSIS IN BLOCKCHAIN

In which the data flow in architecture of blockchain is shown via the administration management system in an associated network of administration worldwide is shown which consist of events that are linked among them and functioned, which usually shows the flow of the data in the in a network of administration through the blockchain for transparency and enhancing the administrative process.

Figure 4.2 shows the diagram of blockchain in the administration. The flow of the procedures through the system in an organization is shown, which consists of procedures that are interconnected among them and worked and linked in the organization, which generally shows the flow of the data in the organization through the various processes. As the data stored in the blockchain is immutable, the path taken by the blockchain for data flow from generation to the end is negligible. For a better flow

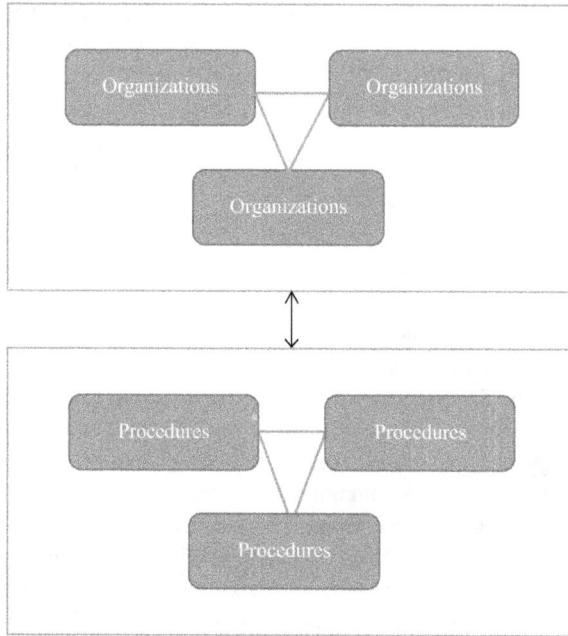

FIGURE 4.2 Data flow analysis.

diagram, the necessary features of blockchain are present. This refers to flow of the digital data acquisition in the internal and external databases for the production data transfer (Khari et al. 2019). The actual databases after preprocessing and the feedback database is not that much necessary, because the infrastructure in each scenario is same so, it can be assumed or neglected (Le et al. 2019). These production technologies are useful and helpful in the blockchain technical production. While looking at the production in the digitization, the chain of a procedural approach is to be followed as:

(1) System
(2) Sensor
(3) Gateway
(4) Internet

A government that is based on the blockchain can safeguard data, lessen fraud, and streamline processes, waste, and the misuse while concurrently increasing confidence and responsibility. A government that is based on the blockchain model government and independent organization businesses divide assets on top of a disperse ledger safeguard by using the cryptography technique. For protecting the sensitive citizen information and the government data the structure given above is used that will also eliminate the single point failures. A government that is based on the blockchain has the capability to resolve provision main points and allow the following advantages:

(1) Safeguard storage of sensitive citizen, government, and the business data
(2) Depletion of labor concerted processes
(3) Depletion of immoderate costs related with managing liability and accountability
(4) Reduced prospective for abuse and corruption
(5) Improve faith in the online civil systems and government
(6) Building up the faith with the citizens
(7) Enhancing efficiency
(8) Shield the sensitive and important data

The administrator-dispersed ledger style can be grasped to assist a cluster of government and public sector processes, which includes the digital payments, identification management, healthcare, voting, supply chain management, and land registration. There are number of governments join the race to process statutory legislation and begin pilot projects that is centered on the blockchain technology. It can leverage the blockchain technology to supply process optimization, cyber security. Government and enterprises have a complex and ever evolving issue of identity and security. For solving the issues related to the digital system and identity, blockchain is used because it has exceptional utility to provide the solutions. There are number of advantages of using blockchain technologies by the enterprises and organizations.

(1) The business will be better protected with the high level of security
(2) The transactions are transparent so they can easily track
(3) Transactions done by the organizations are faster by using the blockchain
(4) The chances of hacking threats are also getting reduced to a larger extent.
(5) There is no need to pay the centralized services because blockchain has the decentralized platforms
(6) Different levels of accessibility are also offered by the organization's blockchain technology
(7) Automatic account reconciliation

4.6   RESEARCH IMPLICATION

In the authors discussed in this paper there are various limitations or gaps in the literature which are as follows. In the existing research of blockchain in various management systems there is a limited number of transactions that are processed in which the number of blocks in the blockchain are restricted in size and frequency. There is a lack of transparency or we can say a lack of centralized transaction processes in the existing blockchain of the management system which leads to a lack of trust in the management systems by people. In the various management system, there is the lack of trust in the network which is used in the organizations, institutions etc. which makes the management systems more vulnerable to cyber-attack. In this paper, authors present a significant appraisal of repeatedly welfare of blockchain technology created in related works and talk over with their suggestion for administrative process. There is a need of move from a "technology-driven" to "need-driven approach" where blockchain utilization is made to order to safeguard fit with necessities of managerial methods and in which the managerial method is improved to get advantage from the technology. The framework for governance is created to be a state for getting advancement. On the basis of significant appraisal, authors offer indication for additional research into the considerable advantages of blockchain applications in the digitization of government and participation of blockchain structure and its utilizations to fulfill community needs of public value. The future work includes the integration of multiple hospital data for early diagnosis of the disease. To work on a combination of a decentralized system and a resilient distributed data approach with cloud computing to invent a health tracking system in broadways.

4.7 RESEARCH LIMITATIONS

The dodges of research met by researchers for putting of suggested and improved administration system in blockchain are labeled as:

(1) The rising, and vast study approved in this context with similar approaches retain out researcher's new approach.
(2) The absence of convenience of credential makes it feasible for explores that definite investigates are inattentive from the research paper as there present some articulators that cannot be openly attainable.
(3) The growing and massive demand of studies in the information security arena generate it probability for researchers to absent off some of the primary seeks from a consistent study directed by investigators

4.8 CONCLUSION AND FUTURE SCOPE

The primary demand focused in the study how blockchain technology is helping in invention and renovation of government process such as application in reducing of corruption, and in health care etc. In this paper, authors present a significant appraisal of repeatedly welfare of blockchain technology established in related works and considered their suggestion for administrative process. There is a need for a move from a "technology-driven" to a "need-driven approach" where blockchain utilization is made to order to safeguard fit with necessities of managerial methods and in which the managerial method is improved to get advantage from the technology. The framework for governance is created to be a state for getting advancement. On the basis of significant appraisal, authors offer indication for additional search into hefty advantages of blockchain application in digitization of government and participation of blockchain structure and its utilizations to fulfill community needs of public value.

- The future work includes the integration of multiple hospital data for early diagnosis of the disease.
- To work on a combination of a decentralized system and a resilient distributed data approach with cloud computing to invent a health tracking system in broadways.

REFERENCES

Bell, L., Buchanan, W. J., Cameron, J., & Lo, O. (2018). Applications of blockchain within healthcare. *Blockchain in Healthcare Today*, *1*(8). 10.30953/bhty.v1.8

Casino, F., Dasaklis, T.K., & Patsakis, C. (2019). A systematic literature review of blockchain-based applications: current status, classification and open issues. *Telematics and Informatics*, *36*, 55–81.

Dimitrov, D. V. (2019). Blockchain applications for healthcare data management. *Healthcare Informatics Research*, *25*(1), 51.

Farouk, A., Alahmadi, A., Ghose, S., & Mashatan, A. (2020). Blockchain platform for industrial healthcare: Vision and future opportunities. *Computer Communications*, *154*, 223–235.

Gökalp, E., Gökalp, M. O., Çoban, S., & Eren, P. E. (2018, September). Analysing opportunities and challenges of integrated blockchain technologies in healthcare. In *Eurosymposium on systems analysis and design* (pp. 174–183). Springer, Cham.

Grover, P., Kar, A. K., & Janssen, M. (2019). Diffusion of blockchain technology: Insights from academic literature and social media analytics. *Journal of Enterprise Information Management 32*(5): 735–757.

Hoy, M. B. (2017). An introduction to the blockchain and its implications for libraries and medicine. *Medical Reference Services Quarterly*, *36*(3), 273–279.

Jha, S., Kumar, R., Chatterjee, J. M., & Khari, M. (2019). Collaborative handshaking approaches between internet of computing and internet of things towards a smart world: a review from 2009–2017. *Telecommunication Systems*, *70*(4), 617–634.

Khari, M., Garg, A. K., Gandomi, A. H., Gupta, R., Patan, R., & Balusamy, B. (2019). Securing data in Internet of Things (IoT) using cryptography and steganography techniques. *IEEE Transactions on Systems, Man, and Cybernetics: Systems*, *50*(1), 73–80.

Le, D., Kumar, R., Mishra, B. K., Khari, M., & Chatterjee, J. M. (2019). *Cyber security in parallel and distributed computing*. Wiley, Hoboken.

Myeong, S., & Jung, Y. (2019). Administrative reforms in the fourth industrial revolution: The case of blockchain use. *Sustainability*, *11*(14), 3971.

Paik, H. Y., Xu, X., Bandara, H. D., Lee, S. U., & Lo, S. K. (2019). Analysis of data management in blockchain-based systems: From architecture to governance. *IEEE Access*, *7*, 186091–186107.

Vos, Jacob, Lemmen, Christiaan, & Beentjes, Bert (2017, March). Blockchain based land administration feasible, illusory or a panacea. In *Netherlands Cadastre, Land Registry and Mapping Agency. Paper prepared for presentation at the 2017 World Bank Conference on Land and Poverty*. The World Bank, Washington, DC.

5 Combination of Convolutional Neural Networks and Blockchain in Handwritten Digit Recognition

Mohit Dayal, Ameya Chawla, Deepti Singh, and Rupesh Kumar Garg

CONTENTS

5.1 INTRODUCTION TO CONVOLUTIONAL NEURAL NETWORKS

Convolution mathematically means we produce a new function by using two functions the same way. In a convolutional neural network we try to convert the input image to a more insightful image where features can be detected very easily by applying many filters on the image to detect the features and have a better training phase of the model. Convolutional neural networks are widely used for problems where images are the input dataset and main objective of the model is to find different patters or features in the images.

Convolutional neural networks are part of deep learning which is a subset of machine learning where we try to replicate the back part of the human brain which is the cerebral cortex which has an important part, the visual cortex, which is used for human vision. Convolutional neural networks are one of the most used machine learning models for image classification, pattern detection, object detection and object recognition due to its structure of filters used to gain meaningful insights from the images. One of the most popular applications are self-driving vehicles which uses object detection algorithm to detect and recognize objects and take decisions on how to move the car.

Blockchain is an expanding record list which are termed blocks and the linking of these blocks is done using cryptography which forms chain-like structures which give it its name blockchain. Each block which is added contains following information like hash code, a timestamp, and transaction data all these are stored in the block and hash code is a cryptographic code which is of the last block which is added to the chain. This type of structure cannot be altered as the block is added for each transaction and it is designed in a way to be effective against modification. The most common use of blockchain technology is cryptocurrency like bitcoin.

Blockchain enables data systems which can be either used in healthcare or student education sector; all can be implemented using blockcahin technology which makes it easier to add user data and is more secure. Then a neural network can be trained on that type of data to make more easier method to analyze the data fit model on it rather than generic way to collect data [1].

Objective: This chapter will give a brief overview of convolutional neural networks to the reader with full understanding about the terms related to the convolutional neural network, architecture of the convolutional neural network and mechanism of propagation of image through various filters and how patterns are recognized from the image. There will be a brief overview of blockchain technology and real-life application of blockchain technology and convolutional neural networks.

5.2   ARCHITECTURE OF A CONVOLUTIONAL NEURAL NETWORK

A convolutional neural network has convolution layers followed by pooling layers and at end fed into a fully connected neural network.

5.2.1   LAYERS

Convolutional neural networks have basically three types of layers:

5.2.1.1   Convolution Layer

Convolution layer has a defined map which moves sequentially over the image and multiply its terms with terms on which it is placed and then product is stored in the new output matrix. The movement of the filter is also defined by value of stride which tells how much blocks the filter move right and down while applying filter over an image. For example, if a filter is placed on top left corner of the image then the filter will move right by one block until it covers the whole breadth of the image then starts

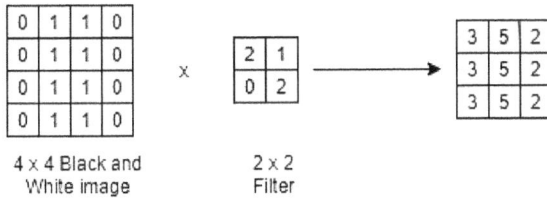

4 × 4 Black and   2 × 2
White image   Filter

FIGURE 5.1   Filter of shape 2 × 2 moved over 4 × 4 image with 0 padding and stride =1.

again from one row down from left side until breadth is covered this is done till the whole image is covered this is and example of filter with stride value as 1. The main purpose of the convolution layer is to detect the features from the image given as input (Figure 5.1).

   We can derive formula for output image size if padding is zero and output image size for a $n \times n$ image is

$$size = n - s + 1$$

in which $n \times n$ is shape of input image and $k \times k$ is shape of convolution matrix.

5.2.1.2   Pooling Layer

The pooling layer works on the same mechanism as the convolution layer but with some small changes like instead of multiplying with terms of the filter the average or max value of the part of the image is stored in the output matrix. Max pooling layer helps to remove the extra information and extract those features from image and removing the extra part left. Multiple times convolution layer is applied with pooling layer after it to extract the most information, features and patterns from the image (Figure 5.2).

   We can derive formula for output image size if padding is zero and output image size for a $n \times n$ image is

$$size = n - s + 1$$

in which $n \times n$ is shape of input image and $k \times k$ is shape of max pooling matrix.

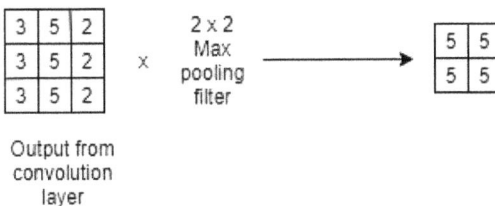

Output from
convolution
layer

FIGURE 5.2   Max pooling layer of shape 2 × 2 moved over 3 × 3 image with 0 padding and stride =1.

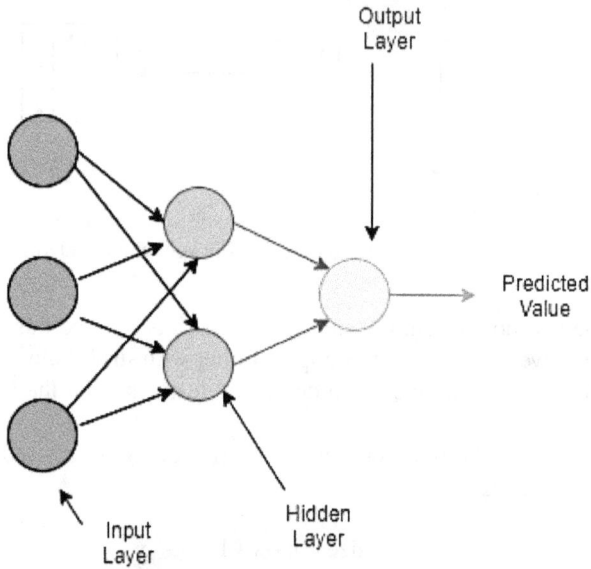

FIGURE 5.3 Example of fully connected layer.

5.2.1.3 Fully Connected Layer

The application of multiple layers of convolution followed by pooling gives a final image with most features extracted. Then that image is converted to a flattened vector or matrix with one column and given as input to a fully connected neural network structure or artificial neural network and then forward and backward propagation is used to train and here the classification of images happens on the basis of its features (Figure 5.3).

5.2.2 PADDING

Padding is added to an image before applying any operation on the image as applying any operation will eventually reduce the size of the image and to maintain a same size of image extra layer of padding is added which make a whole boundary around the image and added boundary values can be either 0 showing the boundary is white or same to pixel on border of the image. Values of boundary depends on the image dataset given as input and padding value n refers to increase in number of rows by 2 × n and columns by 2 × n and the extra values are either white or same as that of the boundary (Figure 5.4).

We can derive equation for output image size which has padding greater than 0 for max pooling layer and convolution layer of size k

$$\text{size} = (n - k + 2p) \ s + 1$$

where $n \times n$ is shape of input image, $k \times k$ is shape of convolution matrix, p is size of padding and s is stride value.

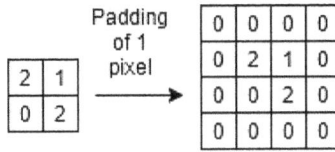

FIGURE 5.4 Example of padding with 1 pixel size.

5.2.3 MIN-MAX SCALING

Min-max scaling is one of the most common techniques used to convert the values of the pixels given in the image to a range of [0,1] to increase the speed and reduce the calculations done by the convolutional neural network and it can be achieved by dividing all values by 255 as 255 is the max value of a pixel of image as pixel can be of one byte and max value stored by one byte is 255.

5.3 WORKING OF A CONVOLUTIONAL NEURAL NETWORK

When we train a convolutional neural network the data is transmitted in each epoch twice through the model once transmitted forward and once backward. The transmission of data through the whole model is called propagation.

5.3.1 FORWARD PROPAGATION

During forward propagation the image is transported from multiple layers of where the filters are made with random values, the same with biases and weights of fully connected layer. On the basis of the output after the forward propagation of the image the output value is used to update all the weights, biases and filter values (Figure 5.5).

5.3.2 BACKWARD PROPAGATION

Backward propagation is the opposite travel direction where on the basis of output an error is generated and hence used to update the weights and biases of the fully connected layer and the filter values. Gradient descent algorithm is used to update all the weights,biases and filter values.

Let's consider an example in Figure 5.6:

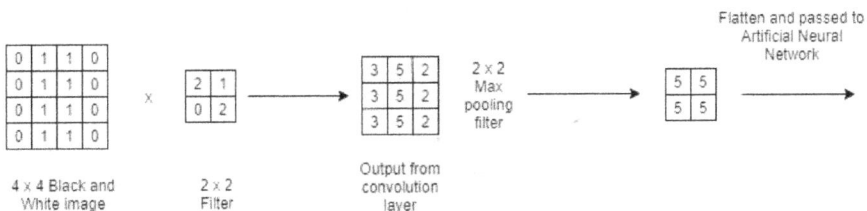

FIGURE 5.5 Propagation of image through various layers.

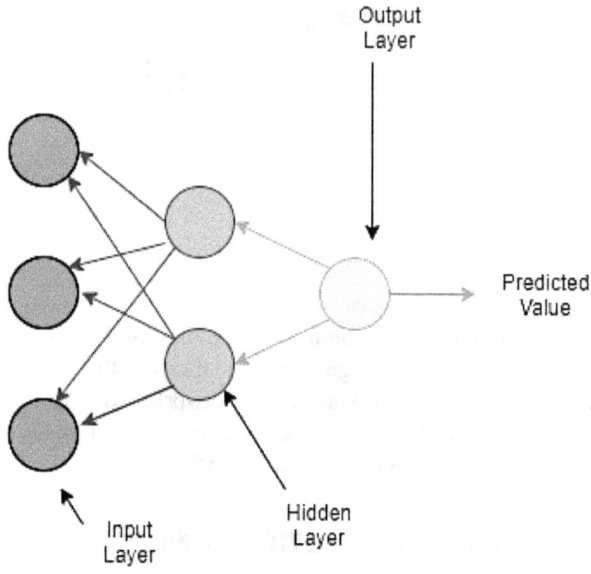

FIGURE 5.6 Backward propagation.

Blue arrows show how data is transmitted from the output layer to the hidden layer with pink-colored neurons. Let's find the loss function which is defined as square of difference in actual value and predicted value.

$$E = \left\{ z_p - z_a \right\}^2$$

z_p is the value predicted by the neural network and z_a is the actual value expected at the output, where z_a is a constant and z_p is calculated using the activation formula in last neuron.

$$z_p = \text{Activation}\left(y_p \right)$$

$$y_p = \Sigma w_i x_i + b$$

Now let's consider the weights in the next Figure 5.7:

Let's consider the output neuron (blue neuron): it is receiving 2 weights W_4 and W_3 and has bias value b_3 so while back propagating the values of them will be updated as:

$$W_4 = \left\{ W_4 - \eta * \text{gradient} \left(\text{loss} \right) \right\} \text{ gradient with respect to } W_4$$

$$W_3 = \left\{ W_3 - \eta * \text{gradient} \left(\text{loss} \right) \right\} \text{ gradient with respect to } W_3$$

$$b_3 = \left\{ b_3 - \eta * \text{gradient} \left(\text{loss} \right) \right\} \text{ gradient with respect to } b_3$$

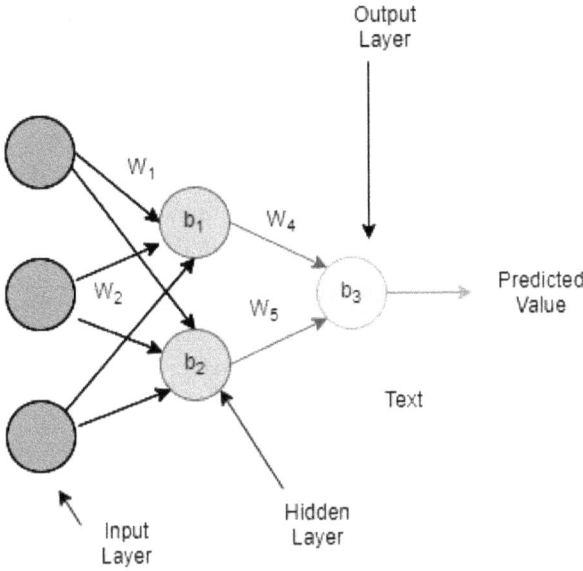

FIGURE 5.7   Backward propagation.

where η is the learning factor which is a hyper parameter and the goal of the algorithm is to update weights and bias for each of the neurons and to optimize the error.

Gradient of loss with respect to W_4 is calculated as:

$$\frac{\partial E}{\partial W_4} = \frac{\partial \{y_p - y_a\}^2}{\partial W_4}$$

$$\frac{\partial \{y_p - y_a\}^2}{\partial W_4} = 2 * \{y_p - y_a\} * \frac{\partial y_p}{\partial W_4}$$

Let's consider linear activation function:

$$y_p = a * \left(W_4 x_4 + W_5 x_5 + b_3 \right)$$

$$\frac{\partial y_p}{\partial W_4} = a x_4$$

W_3, x_4, x_3, b_3 are considered constants while doing partial derivative with W_4. Similarly we can calculate gradient of error due to bias.

$$\frac{\partial y_p}{\partial b_3} = a$$

Let's consider the example of a pink neuron with bias b_2. Let's consider it is sending z_2 as input then:

$$\frac{\partial E}{\partial b_2} = \frac{\partial E}{\partial z_2} * \frac{\partial z_2}{\partial b_2}$$

z_2 can be considered as x_5 sent as input which implies

$$\frac{\partial z_2}{\partial b_2} = a$$

$$\frac{\partial E}{\partial x_5} = \frac{\partial \{y_p - y_a\}^2}{\partial x_5}$$

$$\frac{\partial \{y_p - y_a\}^2}{\partial x_5} = 2 * \{y_p - y_a\} * \frac{\partial y_p}{\partial x_5}$$

As we have already calculated y_p then substituting it we can calculate following gradient.

$$y_p = a * \left(W_4 x_4 + W_5 x_5 + b_3\right)$$

$$\frac{\partial y_p}{\partial W_4} = W_5$$

using the above result

$$\frac{\partial E}{\partial b_2} = a W_4$$

Hence we can calculate gradients for all weights and biases of each and every neuron and update them in each epoch.

5.4   IMPLEMENTATION OF CONVOLUTIONAL NEURAL NETWORK USING PYTHON

There are many ways to implement a convolutional neural network either by making our own class and defining all weights, biases, activation functions and filters, but we will be using predefined modules in Python. We will be using Keras library to create our convolutional neural network module. Data which we will use in this implementation is MNIST dataset which is a collection of handwritten digits.Many fields have practical implementation of CNN [2–4] (Figures 5.8–5.15).

```
#used to create layers for the model and to load the dataset
import tensorflow.keras as keras

#used to plot sample images given in the dataset
import matplotlib.pyplot as plt

#Loading the mnist dataset
from keras mnist=keras.datasets.mnist (x_train,y_train),(x_test,y_test)=mnist.load_data()

#Min Max scaling the pixel values
x_train=x_train/255.0 x_test=x_test/255.0

#Printing the shape of the data set
print(x_train.shape,x_test.shape)
```

FIGURE 5.8 Importing and loading all the required dataset.

```
print(x_train.shape,x_test.shape)

(60000, 28, 28) (10000, 28, 28)
```

FIGURE 5.9 Using method shape for getting shape of the images.

```
#Plotting sample image
plt.imshow(x_train[0].reshape(28,28))
```

FIGURE 5.10 Using imshow method to plot the image of 5.

```
<matplotlib.image.AxesImage at 0x7f4e4a0aaed0>
```

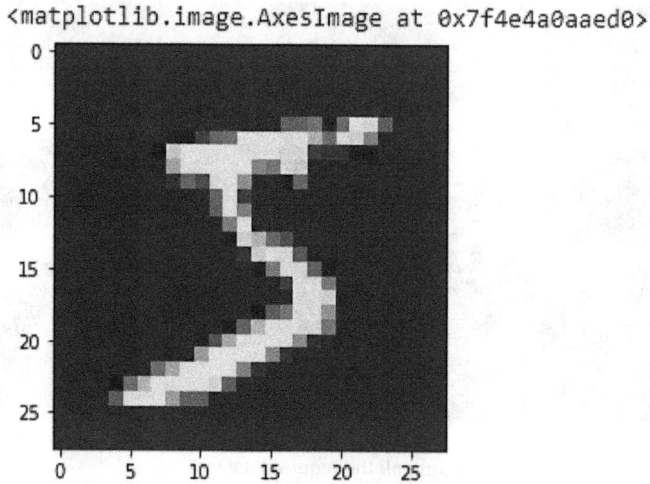

FIGURE 5.11 Using imshow method for plotting the image of 5.

```
#Reshaping the images
    x_train=x_train.reshape((60000,28,28,1)) x_test=x_test.reshape((10000,28,28,1))

#Declaring the input layer
i=keras.layers.Input(shape=(28,28,1))

#Declaring multiple convolution layers and maxpooling layers
#Convolution layers with kernal/filter size 3 x 3 #Max pooling layers with kernal/filer size 2 x 2
c=keras.layers.Conv2D(32,(3,3),padding='valid')(i)
m=keras.layers.MaxPooling2D((2,2),(2,2))(c)
c=keras.layers.Conv2D(64,(3,3),padding='valid')(m)
m=keras.layers.MaxPooling2D((2,2),(2,2))(c) c=keras.layers.Conv2D(128,(3,3),padding='valid')(m)
m=keras.layers.MaxPooling2D((2,2),(2,2))(c)

#Declaring the image flattening layer
f=keras.layers.Flatten()(m)

#Declaring the dense layer with softmax activation
d=keras.layers.Dense(10,activation=tf.nn.softmax)(f)

#Declaring the model with input and output
model=keras.Model(i,d) model.summary()
```

FIGURE 5.12 Initialization of the model and adding all the required layers.

```
Model: "model_2"

_____
Layer (type)                  Output Shape              Param #
=================================================================
input_3 (InputLayer)          [(None, 28, 28, 1)]       0

conv2d_5 (Conv2D)             (None, 26, 26, 32)        320

max_pooling2d_5 (MaxPooling2  (None, 13, 13, 32)        0

conv2d_6 (Conv2D)             (None, 11, 11, 64)        18496

max_pooling2d_6 (MaxPooling2  (None, 5, 5, 64)          0

conv2d_7 (Conv2D)             (None, 3, 3, 128)         73856

max_pooling2d_7 (MaxPooling2  (None, 1, 1, 128)         0

flatten_2 (Flatten)           (None, 128)               0

dense_2 (Dense)               (None, 10)                1290
=================================================================
Total params: 93,962
Trainable params: 93,962
Non-trainable params: 0
```

FIGURE 5.13   Summary of the model showing all layers.

```
#Compiling the model
model.compile(optimizer='adam',loss='sparse_categorical_crossentropy',metrics=['accuracy'])

#Fitting the model on the training data
model.fit(x_train,y_train,epochs=5)

#Testing model for accuracy
test_loss,test_accuracy=model.evaluate(x_test,y_test)
print(test_loss,test_accuracy)
```

FIGURE 5.14   Compiling the model and testing its accuracy.

```
print(test_loss,test_accuracy)
```

0.05564744397997856 0.9836999773979187

FIGURE 5.15 Test accuracy of the model.

The model obtained an accuracy of approximately 98.4% and this accuracy can be
further increased by changing hyper parameters like adding more convolution layers
and max pooling layers and epochs while training.

5.5   LITERATURE REVIEW

1. **Securing CNN Model and Biometric Template using Blockchain**
   Blockchain is one of the most secure systems and recently blockchain has
   been used to convert any normal deep learning model into a secure system.
   Biometric recognition model trained to leverage on blockchain technology
   to provide fault tolerance [4, 5, 6].

2. **An Advanced CNN-LSTM Model for Cryptocurrency Forecasting**
   Cryptocurrency has become one of the most used currencies for transactions and has taken a large part of market share. Most people have started
   investing in cryptocurrency and there is need of a system to forecast the
   value of cryptocurrency so the model was made on past cryptocurrency
   prices of three cryptocurrencies: Bitcoin, Ripple and Etherium [7].

3. **CNN-based Multivariate Data Analysis for Bitcoin Trend Prediction**
   Bitcoin is one of the biggest applications of blockchain and the most used
   cryptocurrency for transactions. Prices of bitcoin are highly volatile and it
   changes the most times due to the number of investors and people using it.
   There is need of forecasting of the price of bitcoin in advance by the model
   made by using 1D CNN [8].

4. **Tree-based Convolutional Neural Networks for Object Classification in
   Segmented Satellite Images**
   Images taken from satellite are always of high resolution which increases
   the cost of computation and processing cost of those images. There is a need
   for an image system where cost of classification, computation and processing is low on these high quality images which can be achieved by using a
   tree-based convolutional neural network [4].

5. **Investigating the Problem of Cryptocurrency Price Prediction: A Deep
   Learning Approach**
   Cryptocurrency has taken over the market share of investment and many
   people have started investing in cryptocurrencies so their is a need of system
   which can make efficient and effective prediction of the price of cryptocurrency to help people to make better investements in cryptocurrency [9].

5.6   OPEN CHALLENGES

There are many open challenges related to the technology of blockchain but the most common challenge is to convince the public to adopt it as most users have a misconception in their mind relating blockchain technology to cryptocurrency which most people think is an illegal currency or currency used by hackers and fraud users. Another main challenge faced in the technology is speed as this system used for transactions is very slow as compared to other transactions systems used in modern-day technologies.

REFERENCES

1 M. Dayal, N. Singh, Indian health care analysis using big data programming tool, *Procedia Computer Science* 89 (2016) 521–527.
2 M. Khari, A. K. Garg, R. G. Crespo, E. Verdu, Gesture recognition of rgb and rgb-d static images using convolutional neural networks, *International Journal of Interactive Multimedia & Artificial Intelligence* 5 (7) (2019) 22–27.
3 M. Dua, R. Gupta, M. Khari, R. G. Crespo, Biometric iris recognition using radial basis function neural network, *Soft Computing* 23 (22) (2019) 11801–11815.
4 Y. H. Robinson, S. Vimal, M. Khari, F. C. L. Hernandez, R. G. Crespo, Tree-based convolutional neural networks for object classification in segmented satellite images, *The International Journal of High Performance Computing Applications* (2020) doi:1094342020945026.
5 A. Goel, A. Agarwal, M. Vatsa, R. Singh, N. Ratha, *Securing CNN model and biometric template using blockchain*, in: *2019 IEEE 10th International Conference on Biometrics Theory, Applications and Systems (BTAS)*, 2019, pp. 1–7. doi:10.1109/BTAS46853.2019.9185999.
6 M. Dayal, B. Nagpal, A compendious investigation of android malware family, *International Journal of Information Privacy, Security and Integrity* 2 (4) (2016) 330–352.
7 I. E. Livieris, N. Kiriakidou, S. Stavroyiannis, P. Pintelas, An advanced CNN-LSTM model for cryptocurrency forecasting, *Electronics* 10 (3) (2021) 287.
8 S. Cavalli, M. Amoretti, CNN-based multivariate data analysis for bitcoin trend prediction, *Applied Soft Computing* 101 (2021) 107065.
9 E. Pintelas, I. E. Livieris, S. Stavroyiannis, T. Kotsilieris, P. Pintelas, *Investigating the problem of cryptocurrency price prediction: A deep learning approach*, in: *IFIP International Conference on Artificial Intelligence Applications and Innovations*, Springer, 2020, pp. 99–110.

6 How Blockchain Technology Can Transfigure the Indian Agriculture Sector
A Review

Urvashi Sugandh, Manju Khari, and
Swati Nigam

CONTENTS

6.1   INTRODUCTION

Blockchain technology is the most talked about research topic from some last years. It has been more than 11 years since the release of the white paper "Bitcoin: A Peer-to-Peer Cash System" by 'Satoshi Nakamoto' in 2008 [1]. The author has given an application of blockchain technology in the area of currency to make it validated cashless. By using the same white paper as a base, various fields have been discovered where blockchain technology can be implemented. And still day-by-day new fields are going to be explored where it could be implemented. Because of the immutable property of the blockchain technology, it has been implemented in many fields to resolve the challenges of that field; some of the examples are: healthcare, the banking sector (Tschorch & Scheuermann 2016), airlines (British Airways), enterprises (Walmart), legal billing, and taxation, Also, recently ideas are being shared for improvement in the agriculture sector [2].

The agriculture sector is one of the vital sectors in the whole world. It plays a significant role from the perspective of business and also to fulfill the need of food for countries. Many countries boost their economy on the basis of their agriculture

DOI: 10.1201/9781003107507-6

productivity. Also, this sector is very dominant as it provides security, nourishment, nutrition, and good health to a nation. For this reason, it is quite important to increase quantity of yields along with the quality of agricultural product. It is possible to achieve by replacing the traditional method to modern methods are linked with the latest technologies. Currently, many countries are using various latest technologies, such as Internet of Things (IoT), machine learning, deep learning, and blockchain, to improve their agriculture sector [3].

Due to the distributed ledger system and immutable properties of blockchain technology, it becomes so popular. Number of systematic literature reviews have been published in the last years. These literature reviews are also helping us for the identification of the topics relevant to agriculture sector that have been studied. We should also know about the challenges and the constraints for future work. In Section 6.3, there will be a discussion about the Indian agriculture sector; in Section 6.4 an architecture is proposed regarding the transformation of the agriculture sector, including its possible use cases followed by the benefits and limitations of blockchain technology.

6.2   BLOCKCHAIN TECHNOLOGY: AN OVERVIEW

Blockchain technology comes in consideration after the development of bitcoin (digital currency) by Satoshi Nakamoto (2008). Blockchain technology may be defined as a distributed ledger technology (DLT) which fetches the unalterable time-stamped value of each transaction between untrusted parties in a decentralized and peer-to-peer network. Records stored in the distributed ledger cannot be changed subsequently. Formation and validation of all records are done by all the parties involved in this network on the basis of consensus protocol. Basically, it consists of a number of blocks and one block is connected to the next one and so on by using a hash function, to form a chain-like structure [4]. This structure can be compared to the structure of a linked list.

I. **Types of nodes in blockchain**: Nodes are basically computers or a machine connected to other computers and used to form a blockchain network. Examples of nodes are servers, laptops, and mobile phones. Nodes are used to store and maintain the data. Increasing number of nodes will increase the level of decentralization of the network. There are two types of nodes in a blockchain, normal nodes and miner nodes, as shown in Figure 6.1:
  - **Normal nodes**: Normal nodes are the standard nodes used to store the whole information in their ledger. It creates multiple copy of blockchain ledger and are responsible for distributing the same. This type of node can reject the incoming transaction if they are not valid.
  - **Miner nodes**: Miner nodes are also known as miners. These nodes are responsible for creating new blocks and adding them to a blockchain. It can be done by satisfying schemes and terms.

II. **Types of ledger and blockchain**: Ledger is a database of transactions in which all the records are maintained. Records stored in ledgers cannot be

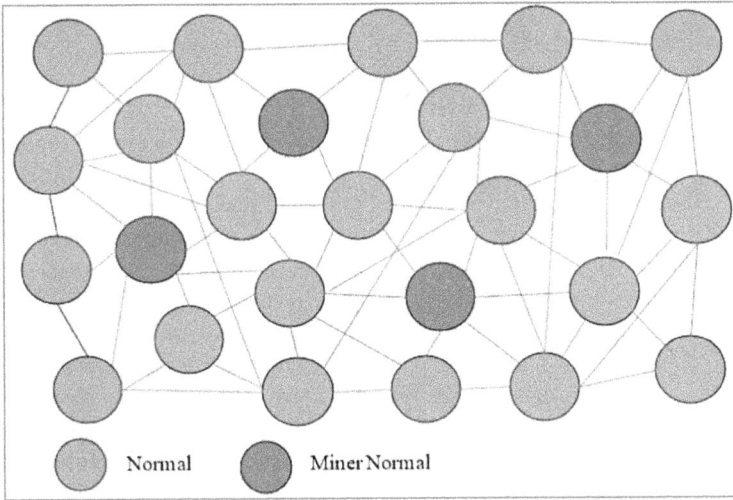

FIGURE 6.1 Example of blockchain network.

manipulated, these are immutable [5]. In blockchain three types of ledger have been introduced: centralized, decentralized, and distributed ledgers, as shown in Figure 6.2. The distributed ledger is further divided into two sub-categories: permissionless (public) and permissioned (private), and these are explained below.

- **Centralized ledger**: In centralized ledger there is centralization of entire data and functions associated with the network. Only the central authority is authorized to take decisions for all the components existing in the same network. Examples of centralized systems are banks, and social media like Instagram, Facebook, and YouTube. A centralized system remains at high risk of system failure and security lapse [6].

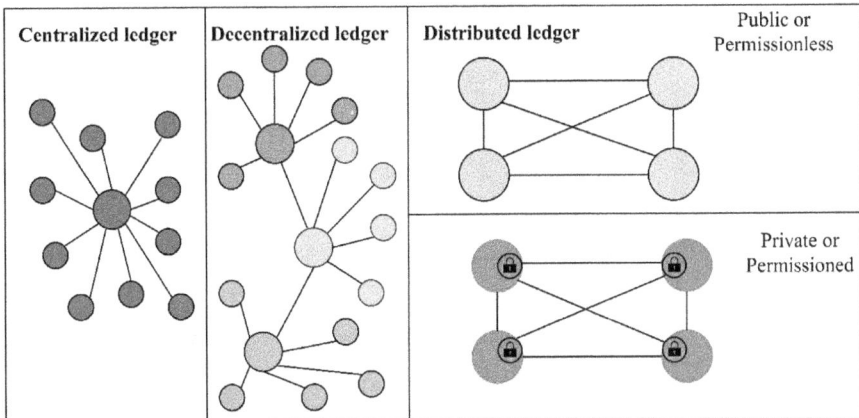

FIGURE 6.2 Different types of ledger.

- **Decentralized ledger**: In this type of ledger, no single authority is responsible for taking charge of the data and all functions of the network. A number of authorities are involved in this system which works for a subgroup of end users. Each user is an independent entity and has its own copy of the ledger. It reflects the changes simultaneously if made in data stored on the ledger.
- **Distributed ledger**: From the control's attitude, nodes of the distributed ledger act as decentralized while they are distributed according to their location. In this network each node acts as a single unit as there are no end users. The distributed ledger is divided into two subcategories:
  - **Public (Permissionless)**: Public ledgers are known as permissionless ledgers. There is no need for permission for the user to join this type of network. A public ledger provides the facility for everyone to make transactions, validate the blocks, etc. without any need for permission. An example of a public ledger or permissionless ledger is bitcoin.
  - **Private (Permissioned)**: Private ledgers are also known as permissioned ledger. The users can join this type of ledger only if they have permission to make transactions. If the user is authenticated, they can use all the facilities associated with the permissioned or private ledger. An example of a private ledger is Hyperledger Fabric.

The first type of blockchain is explained by the bitcoin – that is, a public blockchain – and rest of the types of blockchain are private blockchain and consortium blockchain [7]. A brief description is given in Table 6.1 below:

TABLE 6.1
Differences between Public and Private Blockchain

Parameters	Public Blockchain	Private Blockchain	Consortium Blockchain
Type of user	Permissionless and may be anonymous	Permissioned and known identities only	Permissioned and known identities only
Miners or validators of blocks in blockchain	All peers	One organization	Selected peers
Transaction speed	Slow	Lighter & faster	Lighter & faster
Type of control on network	Decentralized	Centralized	Partially centralized
Cost	Expensive	Less expensive	Less expensive
Immutability	Impossible to tamper with the data	Data could be tampered with	Data could be tampered with
Consensus mechanism	PoW, PoS, PoI, PoB, PBFT	Custom, multi-party	Custom, multi-party

III. **Consensus mechanism**: Consensus mechanism is a protocol which makes an agreement and must be accepted by all the nodes in that network. It helps to validate the transaction occurring between the nodes [8]. The consensus mechanism is important for blockchain network to confirm the connectivity of the nodes to the same network. Various types of consensus mechanism are explained as follows:

- **Proof of Work (PoW)**: This mechanism is used to create new blocks in the blockchain, and it is done by solving the problem. This computational puzzle is assigned to miners as a challenge. In PoW consensus mechanism a lot of energy or computational power is required which makes this mechanism very costly. Example of cryptocurrencies using PoW are Litecoin, Monero coin, and Ethereum.
- **Proof of Capacity (PoC)**: Proof of Capacity consensus mechanism is also known as Proof of Space (PoS). This consensus mechanism is very interesting. In this algorithm, mining nodes use the space of digital storage like hard disks to store the data of challenges offered by the service provider. This is known as plotting. After that, these pre-stored solutions are used to solve the challenges. This PoC is considerably faster than PoW. For this reason, PoC takes far less time to create blocks as compared with PoW.
- **Byzantine Fault Tolerance (BFT)**: This is a property in which the system has the capability to tolerate faults in the network. BFT arranges the records in a proper way which cannot be tampered with until the number of betrayers is more than one-third of the total number of nodes in the network.
- **Practical Byzantine Fault Tolerance (PBFT)**: This was introduced by Barbara Liskov and Miguel Castro. The concept of this mechanism is to bear the inoperative and venomous nodes in a distributed system to acquire consensus. Later, these nodes will be replaced by duplicate nodes when the number of faulty or malicious nodes is greater than one-third of all nodes in the network. An example of the implementation of PBFT is Hyperledger Fabric.
- **Proof of Stake (PoS)**: In this consensus mechanism, nodes are selected randomly to validate the blocks and selected nodes are known as validators. This mechanism is quite different and interesting from the previous consensus mechanism. In this mechanism, there is no chance for the validators to acts as a competitor in respect to the other nodes. Validators stake their cryptocurrency (coins or tokens) to participate in activity of blocks validation in a limited manner. They are selected randomly on the basis of the number of maximum coins to solve the challenge. After validation of the block, validators are rewarded with some proportion of their stake. This process needs less energy, lowers the cost, and is faster in comparison with PoW.
- **Delegated Proof of Stake (DPoS)**: This mechanism is used to select the validators which is not based on the maximum numbers of coins or cryptocurrency. In DPoS, all the nodes having cryptocurrency make a group of delegates whose task is to authorize and validate the transactions. This consensus mechanisms remains decentralized.

- **Proof of Elapsed Time (PoET)**: This mechanism is frequently used in permissioned blockchain networks. PoET is used to make decisions on the mining rights and the winners of the block on the network. The basis of this mechanism to work is on the waiting time. Each node involved on the network is assigned with some waiting time randomly and the node which completes the assigned waiting time is rewarded with the new block. Repetition of the process will be done to discover the new block. This mechanism is an energy saver as nodes go to sleep during the waiting time. PoET is used in Hyperledger Sawtooth and in resource-efficient mining.

IV. **Key components of blockchain networks**: To set up any network, there is a requirement for hardware, software, protocols, etc. Similarly, there are some key components with which the blockchain network can be modeled:
  - *Nodes*: In the blockchain network, computers are considered as nodes. Already, we have discussed the normal and miner nodes.
  - *Miners*: These are the particular nodes whose task is to verify and validate the exchange of information or transaction.
  - **Block**: This is defined as a data structure whose job is to record all the transactions and acts as page of the record file or the ledger.
  - **Chain**: This is formed by an arrangement of blocks in a defined order.
  - **Transaction**: This is the record of all the information exchange between two or more nodes and is further broadcast to the entire network.
  - **Consensus**: This is a protocol or terms accepted by the nodes existing in a network.
  - **Smart Contracts**: Smart contracts are the coded program helps to provide a safe environment for the communication. Also, it examines whether the nodes or devices work in their limits.

V. **Block structure**: The architecture of each block has a block header and a block body as shown in Figure 6.3. Block header has the following fields:
  - **Version**: Size of this field is of 4 bytes and used to have the knowledge about protocol updates.
  - **Time stamp**: This value is used to determine the formation and acceptance of data in that block by the network. it has size of 4 bytes.
  - **Previous Block Hash**: This field is used to store the hash code of previous block. This hash code is responsible to make relationship and maintain the order if its (block) creation with the other block exist in the blockchain. Previous block hash has the size of 256 bytes.
  - **Merkle tree**: This is created by using the hash of each transaction associated with it. The structure of Merkle trees is almost similar to binary tree. In Merkle trees, each hashed transaction takes place in same block, and is again hashed to form a tree. Size of this field is of 32 bytes.
  - **N-bits**: This defines the difficulty target to encountered by the miners to resolve a block.
  - **Nonce**: This is a short form of "number only used once" and is of 4 bytes. Nonce is a pseudo-random number that is further used by the miners for a valid block.

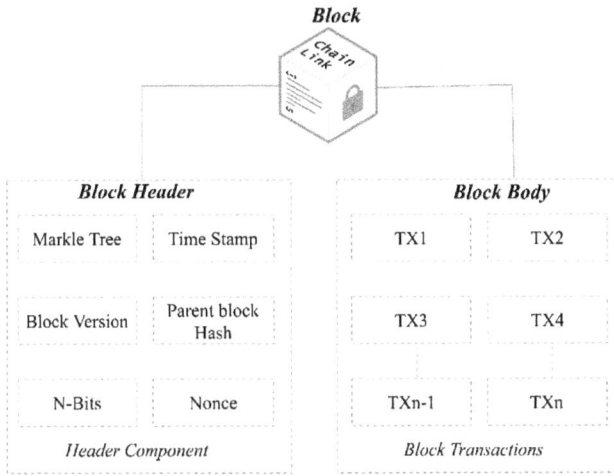

FIGURE 6.3 Structure of a block.

The size of a transaction in a block is approximately 250 bytes. A block header is a field of 80 bytes in a block. The size of a complete block having maximum of its transaction is approx. 1,000 times of the block header. The rest of the fields are as follows:

- **Block size**: Represents the size of block of 4 bytes.
- **Transactions**: It is a variant and stored in the same block.
- **Transaction Counter**: It is variant in size (1–9) bytes and defines how many transactions have been done.

VI. **Block creation and verification**: A blockchain consists of a number of blocks which are used to store the data generated by the number of transactions takes place in the network. To make a fresh transaction, there is a requirement of public key and private keys for the implementation of secure and successful transaction between two parties. During a transaction, both of these keys are used to generate the digital signature which are further used to authorize and validate the transaction [9]. If any third tries to make changes in the transaction data leads to the drastic change in the digital signature. To validate the transaction on the network various consensus algorithm are used. In short, if a node is interested to make transaction, this transaction is further broadcast to every node in a peer-to-peer network as shown in Figure 6.4.

The next step is to validate and authenticate the transaction done with the help of consensus mechanism. After the validation of transaction, it will get added in the block. When a block is reaches to its maximum capacity then it is added to the current blockchain with the help of a hash function [10]. This function is used to generate the address, digital signatures and is used in mapping of data from arbitrary-size

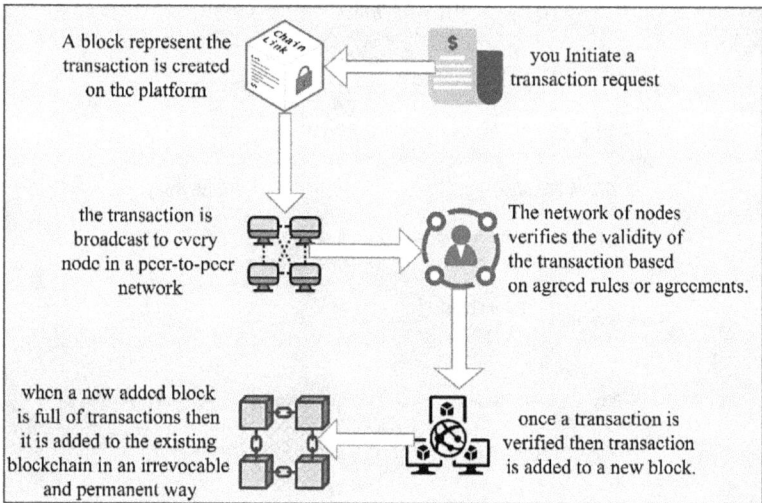

FIGURE 6.4 Working of blockchain.

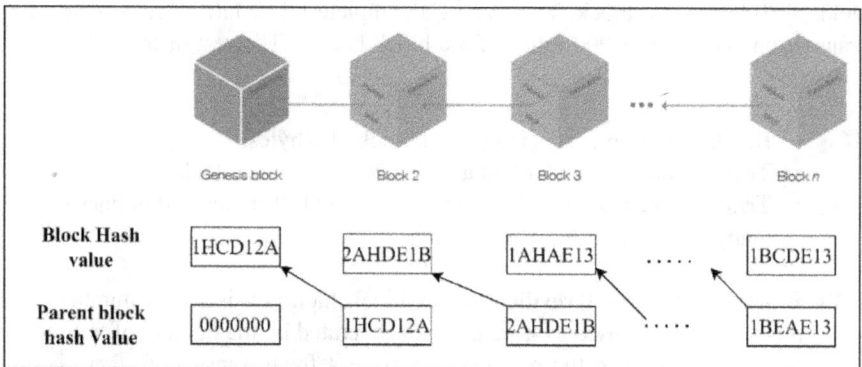

FIGURE 6.5 Blockchain as a sequence of node.

to fixed-size values. Each block is linked to the next one by using hash value of the previous block as "previous block hash value" as shown in Figure 6.5.

6.3 INDIAN AGRICULTURE SECTOR: AN OVERVIEW

I. **The facts**: The Indian agriculture sector is growing by using technologies. In this section we will take an overview of the Indian agriculture sector. This sector is very important to the population of India as around 60% of them survive on farming. The agriculture sector constitutes about 7.8% of total agriculture products of the globe. For this reason, this sector is providing food to 1.42 billion of people from farm to mouth. Export food is done worldwide on a very large scale. On the latest parameters around 52.3% of

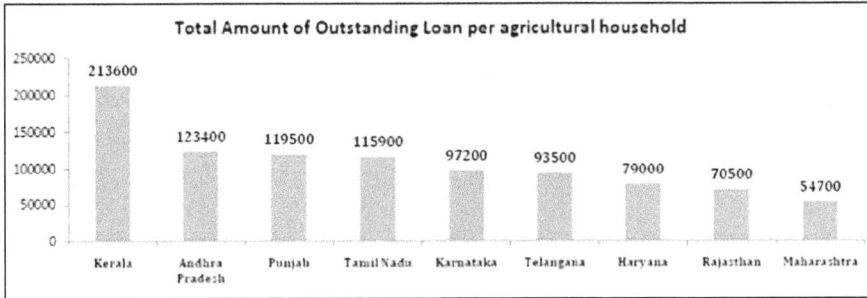

Total Amount of Outstanding Loan per agricultural household

FIGURE 6.6   Statistical data about financial crisis centralized.(Image source: Isha Outreach)

India's population is considering agriculture as their source of earning. Due to the reason, agriculture sector contributes approximately 13.8% to GDP. After having all this, farmers in India have been committing suicide and numbers of farmers are not feeling good and are leaving agriculture.

   II. **The status of Indian farmers**: The farmers, to whom everyone shows a very positive gratitude and feels grateful to them, work so hard to grow and cultivate the crops at their best. Because of their hard work to cultivate good crops at their best. Despite their hard work, they don't have a meal of one time to fill their stomach.

  III. **Main reason for farmers' financial crisis**: There is no financial stability in the agriculture sector as many risk factors are associated with it. Figure 6.6 shows some statis data of financial crisis. The risk factors are:
   • Changing Climate
     – Unpredictable weather
     – Lack of and unequal distribution of rainfall
     – Limited supply of irrigation water
     – Regular failure of yields.
   • Lessen the size of land
     – Unequal partitioning of land
     – Approximately 62% of agricultural land is easily affected by natural disasters – flood or lack of water, etc.
   • Storage facility
     – Due to lack of cold storage, agricultural products are damaged
     – As a result, selling of agricultural products at an inappropriate price in the market.
   • Unprofitable pricing of products
     – Farmers are not able to sell the agriculture products in the market at appropriate price due to lack of knowledge.
     – Brokers or middlemen take advantage of profit and farmers remain with the lower price.
   • Liabilities and their reason
     – Most of the farmers don't use the banking system, in consequence of which they become dependent on private money lenders at higher rate of interest.

- Prerequisites needed for agriculture are quite expensive.
- Support of government is at almost not there due to which farmers are not able to repay their loans.
- Middlemen or brokers buy the agricultural produce at low rates and sell it on at very high prices which makes the farmers poor day by day.
- Recursive failures of crop compelled the farmers to lend the money.

IV. **How the agriculture sector is falling**:

Figure 6.7 shows farm loan statistical data in 2016 which increases continuously. Although 52.3% of India's population survive on the basis of agricultural produce, this sector is lagging behind. They do a lot of hard work to cultivate good crops, but in exchange they receive a very low amount to sustain their living. Because of its unprofitable results people are moving to some other occupation. In India, Farmers are almost dependent to nature like climate, rainfall, drought etc. per profit perspective, farmers are not satisfying with the output of agriculture in terms of profit. They are using the old methods of farming. There is no financial assistance to buy the required input needed before seeding the fields [11]. They need a guidance on farming which is not delivered to farmers in a proper way. In the agriculture sector, there is a need to modify the traditional way of farming. New technologies must be to avoid the falling of the agriculture sector and farmers must know its usage. To avoid the over storage of crops, the system must be traceable so that farmers can pass the middlemen/brokers to sell their products at optimum prices.

V. **Current state of agriculture industry**:

The crisis of Indian agriculture has already been discussed. Problems of the agriculture sector that exist in the current scenario are venturing out. In the

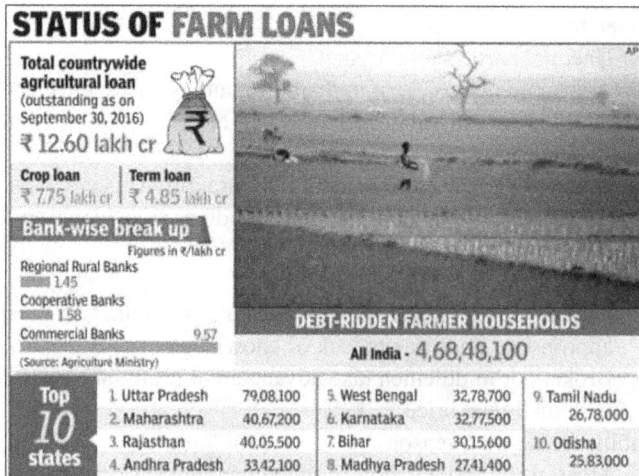

STATUS OF FARM LOANS

Total countrywide agricultural loan (outstanding as on September 30, 2016)
₹ 12.60 lakh cr

Crop loan	Term loan
₹ 7.75 lakh cr	₹ 4.85 lakh cr

Bank-wise break up
Figures in ₹/lakh cr
Regional Rural Banks 1.45
Cooperative Banks 1.58
Commercial Banks 9.57
(Source: Agriculture Ministry)

DEBT-RIDDEN FARMER HOUSEHOLDS
All India - 4,68,48,100

Top 10 states

1. Uttar Pradesh	79,08,100	5. West Bengal	32,78,700	9. Tamil Nadu	26,78,000
2. Maharashtra	40,67,200	6. Karnataka	32,77,500		
3. Rajasthan	40,05,500	7. Bihar	30,15,600	10. Odisha	25,83,000
4. Andhra Pradesh	33,42,100	8. Madhya Pradesh	27,41,400		

FIGURE 6.7   Farm loan statistical data.(Image source: The Economic Times)

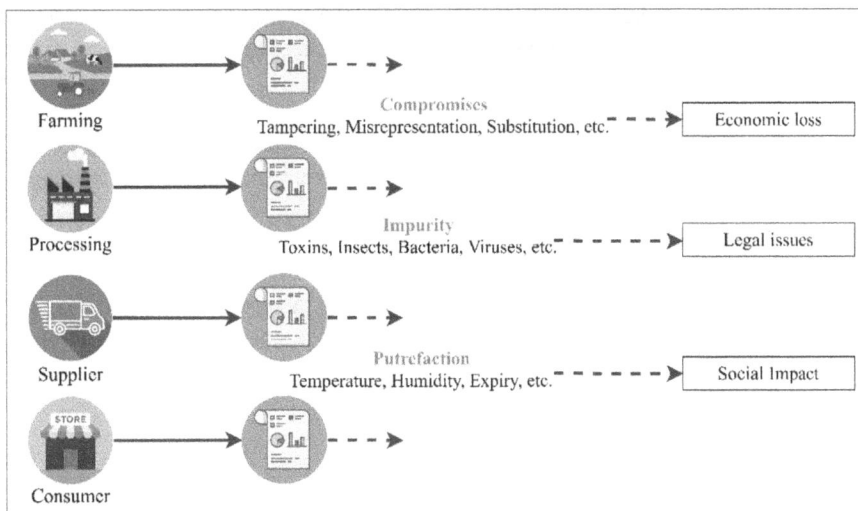

FIGURE 6.8 Current state of agriculture.

current scenario of agriculture industry as shown in Figure 6.8, compromise is made with actual records of agriculture produces in terms of tampering, misrepresentation substitution which leads to economic loss. In the next step produces received from farms are processed where the contamination is done with the use of toxins, bacteria etc. Contamination is done so that the farming products will stay fresh for long and this step may be a legal issue. These contaminated produces are delivered to distributors [12]. Now at this stage, some of the quantity of products get spoiled due to high temperature, Humidity and may be expired. Further, the remainder of the quantity distributed to retailors which is consumed by the consumer. The solution of these problems will result in an increasing rate of production. Due to the reason quality of the product will get increase. But in developing growth rate of population is almost double as compared with the developed countries. Somehow this factor is also a big obstruction in the development of this sector [13].

6.4 HOW BLOCKCHAIN CAN TRANSFIGURE THE AGRICULTURE SECTOR

I. **Proposed architecture**: As we discussed above, blockchain and decentralized ledger gained fame because of its financial application bitcoin, but after exploring, researcher found it has a wide range of different application area outside cryptocurrencies. This technology is all set to extremely transform various industries like, law, banking, healthcare, real estate, etc. Still, one of the least explored sectors that blockchain technology has the potential to transform completely is the Indian agriculture sector. More prominently, it has a large number of challenges that need to be overcome soon. Blockchain technology can help to transform the Indian agriculture sector in many different ways.

IoT, coupled with blockchain technology, is remodeling the agriculture industry. It is all set to make a sustainable farming practice by using data of farming resources. It will use an easy tactic to enhance the use of farming resource like water, fertilizers, labor, transport, etc.

Let us discuss how blockchain technology can facilitate all the stakeholders and growers to transform the agriculture sector. Figure 6.9 shows the proposed architecture of using blockchain in agriculture sector. There are for key section in proposed architecture one is stages of data generation, second is smart contracts, third is blockchain platform, and last one is data consumption.

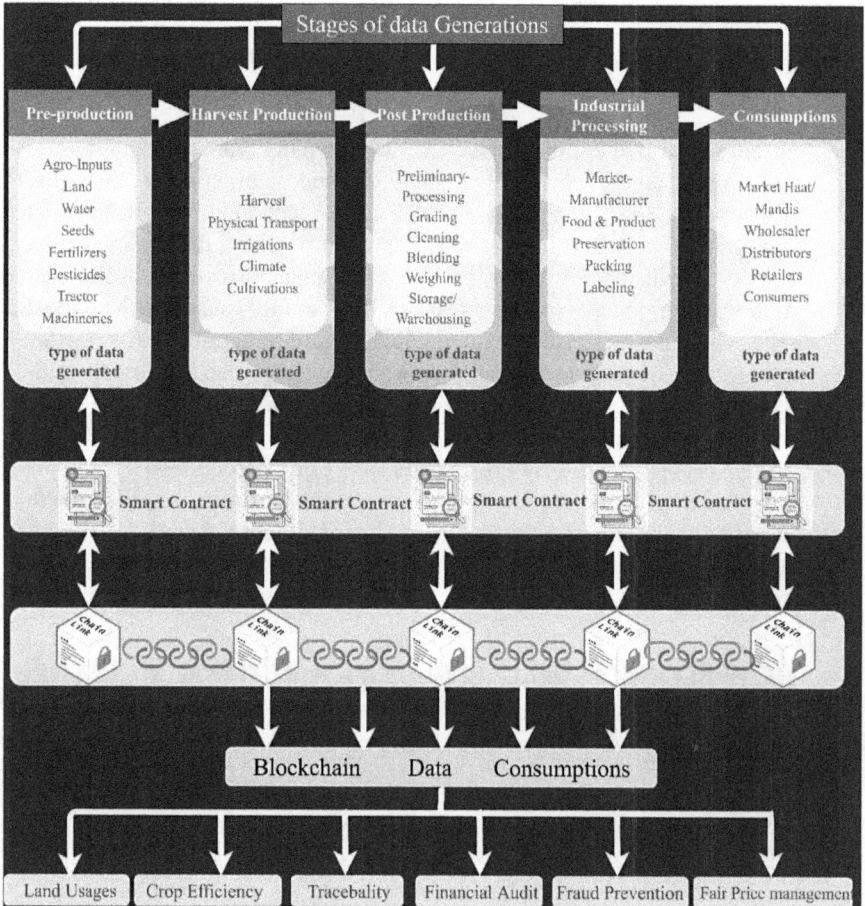

FIGURE 6.9 Proposed architecture.

At the first stage, the system will collect all the data from different phases form various resource. A system is constructed to keep an eye on data like temperature, soil moisture, light, humidity, pH, seeds, fertilizers, and so on, generated in the pre-production phase. Some data like harvest, physical transport, irrigation is generated during the production corp. After production, data about processing, cleaning, blending, storage, and warehousing, etc. is generated which is very useful for consumers to know how food is processed and cleaning. Some vital data is produced during industrial processing, such as packing, expiry date, and details of manufacture.

At the second stage, we use smart contract because smart contracts are automated which means when condition match, they work as a trigger. Smart contract is written to get data from first stage and send to decentralized ledger blockchain. At third stage, we can use a private blockchain like Hyperledger Fabric, or any. All the data generated at the first stage is stored in a blockchain with the help of smart contract. In the end, we can use that data for different purposes, such as land usages, crop efficiency, traceability, and fraud prevention, which the authors will explain in the following sections.

II. Benefits of blockchain in agriculture: Figure 6.10 shows some important benefits of blockchain use in agriculture which explain below:

- *Fraud prevention*: One major feature of blockchain is, once a transaction is complete, data is stored and cannot change. The data stored in blockchain cannot be modified or delete which makes the system protected from any kind of fraud.

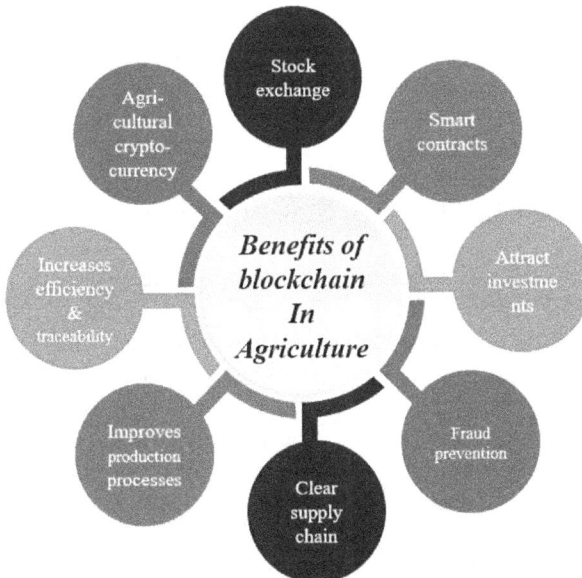

FIGURE 6.10 Benefits of blockchain in agriculture.

- *Clear supply chain*: With the help of shared blockchain ledger, all stakeholders can manage agricultural shipping and monitor the complete process, which improve worth of agronomic goods, because crops, livestock, and machinery can be monitors by the beneficiary to insure ethicality and quality [14]. Consumers can easily trace and verify the journey of product from the farm field to his door because consumer willing to know about the origin of food.
- *Improves production process*: Data is the modern-age oil and is the biggest asset for organization after workforce. All the data collected from sensors and other sources uploaded in the blockchain which help to improve production process [15, 16].
- *Increase efficiency and traceability*: Demand for local agricultural produce and organic produce is increasing daily. With the help of blockchain implementation in agriculture, consumer can easily trace and verify the journey of product from the farm field to his door. Blockchain store all the data about who grow this product, who harvest, who produce and in the end who transport this product [17].
- *Agriculture cryptocurrency*: Bitcoin is the leading cryptocurrency globally. After the huge success of bitcoin, approx. 4,000 cryptocurrencies invented and circulated in market. With the help of blockchain farmers can have their own cryptocurrency or trade in cryptocurrency.
- *Stock exchange*: There are many cryptocurrencies in the market for trading and have a stock exchange for that. If blockchain is used in agriculture and use cryptocurrency, so start-ups can get investment using stock exchange.
- *Smart contracts*: With the help of smart contracts, growers can get their payment on time because smart contracts are automated.
- *Attract investment*: Blockchain implementation in agriculture provides an opportunity for growers to attract funds with the help of ICO or Initial Coin Offering, which is a new way for crowdsource using cryptocurrency.

III. **Limitations of blockchain in agriculture**: While focusing on profitability in the Indian agriculture sector, under unfavorable weather conditions, growers face many challenges. Some of them given below and shown in Figure 6.11:
- *Inaccuracy and complexity of data*: Data is the new weapon and if data is wrong anybody can lose the war. Data collected from sensors or inserted by the people cannot always be trusted. Thus, the collected information may not be consistent. Data collected from agricultural supply chain is very complex [18, 19].
- *Consume too much energy*: The first blockchain platform is bitcoin which use PoW consensus mechanism which depend upon the hard work of miners and miners get some reward to solve a mathematical problem. System use very high energy to solve these mathematical puzzles [20]. After bitcoin, many blockchain framework comes with different consensus algorithms which use less energy but still energy consumption is major problem with blockchain.

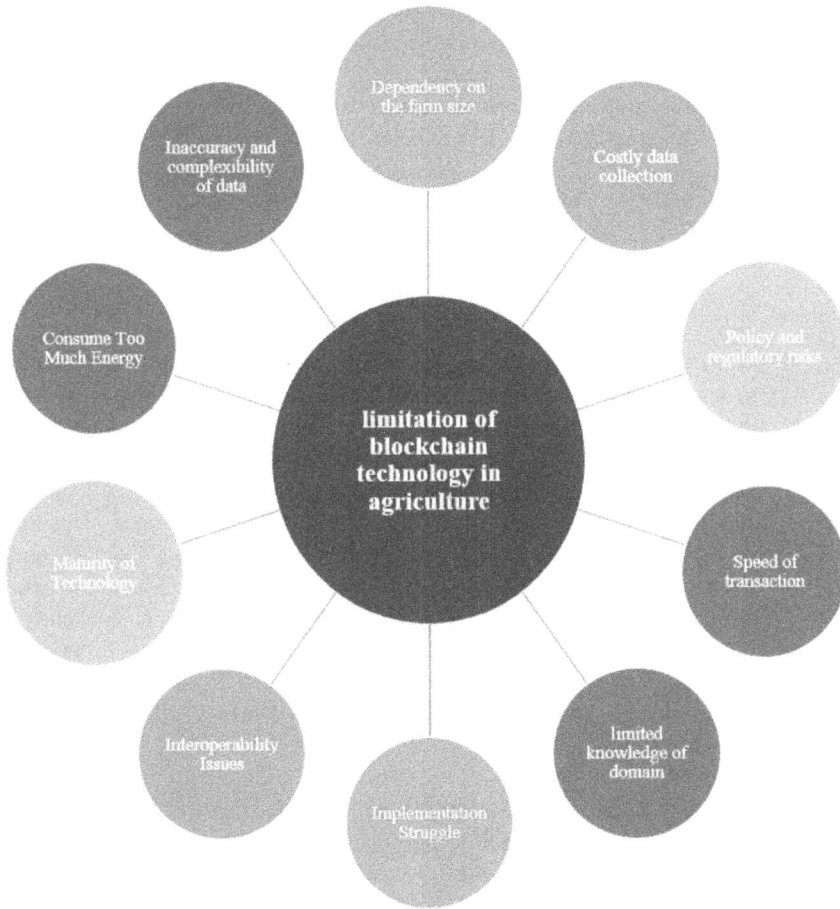

FIGURE 6.11 Limitations of blockchain technology in agriculture.

- *Maturity of technology*: Blockchain technology is only ten years old and not much used in agriculture. This means that blockchain is new technology which needs time to be mature.
- *Interoperability issues*: One major limitation of blockchain is the interoperability issue. In market, there are different types of blockchain platform which work in different way totally and try to solve in their own way. This creates an issue where different blockchain platform cannot communicate effectively [21, 22].
- *Cost and implementation struggle*: Blockchain technology implementation cost is huge, although some platform like Hyperledger is open source, still the need huge investment from enterprise which want to pursue it.
- *Limited knowledge of domain*: Blockchain technology is only ten years old and not much used in agriculture. This means there is a lack of expert

workforce. To implement blockchain in agriculture, they require multiple experts from different domains, and it is always hard to find suitable workforce with domain expertise.

- *Speed of transaction*: Bitcoin blockchain takes one to ten minutes to complete a transaction. Blockchain implementation in agriculture is affected by this transaction speed. Transaction speed plays a vital role when choosing a blockchain platform.
- *Policy and regulatory risks*: One of the major challenges faced by developers in agriculture is country policy and regulation. India has very different policies for agriculture. Indian govt. issue some regulatory for cryptocurrencies also which affect blockchain implementation.
- *Costly data collection*: Data is the modern-age oil and is the biggest asset for organization after workforce. All the data collected from sensors and other sources uploaded in the blockchain and this data collection is very costly because framer need to use many types of sensors. For small farmer it is one of major limitations.
- *Dependency on the farm size*: One of the major challenges of using blockchain in agriculture sector is that sometimes benefits depend upon size of farmland. For example, small growers may fight to integrate and collect farm data, as due to low budget they cannot afford blockchain infrastructure, on the other hand, small farmers can get farm insurance via blockchain.

IV. **The possible use case of blockchain in agriculture**: According to the Indian govt. economic survey 2020–2021, approximately 50% workforce of India depends upon the agriculture and related sector and the share of India agriculture sector in Indian GDP (Gross Domestic Product) is 20% which makes this sector is the leading one in Indian GDP performance. Indian govt. looking forward to increase GDP share of agriculture sector to approx. 30% in next 5 year and want to double the income of Indian farmers. Indian agriculture sector needs technology revolution with the help of new age technology. Blockchain technology will play a major role to restructure agriculture sector in India. There are some areas given below and shown in Figure 6.12 where we can use blockchain technology in agriculture sector to redesign and to boost economy with transparency.

1. *Agri supply chain*: According to a US agriculture department report, the implementation of certain ASM (agricultural management software) will generate as much as 65%–70% higher profits on the same land. These ASM provide complete supply chain management in agriculture sector from growers to consumers. Blockchain technology can play a vital role in this area [23]. With the help of shared blockchain ledger, all stakeholders can manage agricultural shipping and monitor the complete process, which improve worth of agronomic goods, because crops, livestock, and machinery can be monitors by the beneficiary to insure ethicality and quality.

2. *Fair pricing management*: Inflexible demand, climate condition, inelastic supply, and global market conditions all influence the statistics that

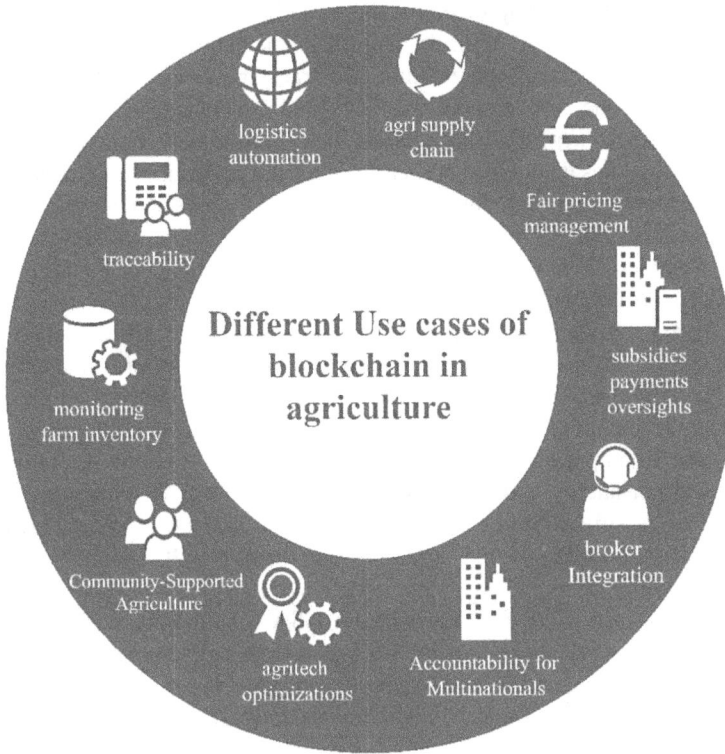

FIGURE 6.12 Different use cases where blockchain can be used in agriculture.

the farmers' income and product prices are tremendously unstable. In a developing country like India middlemen and brokers also play a negative role in farmers' income. Middlemen store the crops and control market pricing, while farmers suffer. With the help of blockchain technology in agriculture, every player can monitor and trace the crop which will reduce stocking the crop at middleman level and can manage price.

3. *Subsidies and payment oversights*: The Indian government provides a wide range of subsidies, such as irrigation subsidy, power subsidy, fertilizer subsidy, agricultural infrastructure subsidy, credit subsidy, agricultural equipment subsidy, seed subsidy, and export subsidy, to farmers but frequently these subsidies from government do not reach the right person. The Indian government implements direct bank transfer scheme to the farmers which is very useful but still there are some problems to trace [24]. With the help of blockchain, government can trace and monitor individual farmer subsidy, which help both the farmer and government.

4. *Broker integration*: In developing countries like India middlemen and brokers are part of the agriculture supply chain. Middlemen and brokers buy the crops from farmers and sell to mandis or corporates with huge margins, resulting in price hikes which affect the consumers. Many

times, with the help of politician, these broker control market price of agricultural goods. With the help of blockchain technology in agriculture, every player can monitor and trace the crop which will reduce stocking the crop at middleman level and can manage price.

5. *Accountability for multinationals*: In the global market, there are always some names in all kinds of sectors which are polarizing, controversial, or argument-inspiring. Some companies cheat farmers and not release payment on time. Sometimes not offer the suitable price of the crops. To overcome from these types of anomalies, blockchain can be used which provide complete transparency [25].

6. *Agri-tech optimization*: According to a US agriculture department report, the implementation of certain ASM (agricultural management software) will generate as much as 65%–70% higher profits on the same land. Technologies like machine learning, artificial intelligence, IoT, cloud computing shows good result in agriculture sector. According to Gartner survey, blockchain technology with artificial intelligence will change agriculture completely. Blockchain will used for transparency and traceability while artificial intelligence will used to analysis everything.

7. *Community-supported agriculture*: CSA (community-supported agriculture) is very popular in the USA, and India is also practicing CSA but in its own manner without technology. CSA bring together a number of persons from community, who helps the farmers financially in return for some sharing of crops or some percentage of income. Blockchain can improve the CSA operation and provide the transparency to all the members of CSA which help to distribute conflict free income as per predefined agreement.

8. *Monitoring farm inventory*: In order to avoid losses after harvesting, growers need to be very active to monitor storage techniques of crop. They must ensure CO_2 absorption under 600 to avoid infestation and mildew growth. Luckily, there are many types of sensors available in the market which play a very vital role to detect probable losses 4–5 weeks prior than old-style temperature monitoring methods. With the help of blockchain technology growers can easily monitor the inventory and offer continuous support to all the key players about what requires to be done.

9. *Traceability of goods*: Demand for local agricultural produce and organic produce is increasing daily. With the help of blockchain implementation in agriculture, consumer can easily trace and verify the journey of product from the farm field to his door. Blockchain store all the data about who grow this product, who harvest, who produce and in the end who transport this product.

10. *Logistics automation*: Blockchain has the potential to record and track every single transaction among the parties. Many organizations are now using blockchain in logistics sector to improve efficiency, for immutable tracking, to enhance security, and for automated smart contracts.

6.5 CONCLUSION AND FUTURE WORK

In this study it has been observed that blockchain technology can maintain the privacy of records as not allowing the altering of data and thus providing a trustworthy environment. This technology can be used for the improvement of the agriculture sector in combination with other technologies such as IoT. In this chapter, we have tried to illustrate the facts about the Indian agricultural industry, detailing the current situation of farmers, and reasons for failure in the same. Also, the authors have proposed an architecture to show the conceptual implementation of blockchain technology. From the perspective of future work, it is possible to discuss some other new issues such as challenges and limitations raised by the recommended solutions.

REFERENCES

1   N. Niknejad, W. Ismail, M. Bahari, R. Hendradi, and A. Z. Salleh, "Mapping the research trends on blockchain technology in food and agriculture industry: A bibliometric analysis," *Environ. Technol. Innov.*, vol. 21, p. 101272, 2021, doi:10.1016/j.eti.2020.101272.
2   P. Mukherjee, R. K. Barik, and C. Pradhan, "Agrochain: Ascending Blockchain Technology Towards Smart Agriculture," in *books.google.com*, 2021, pp. 53–60.
3   S. Mohapatra, K. C. Anirudh, R. K. Nithin, G. Bhandari, and J. Nyika, "Application of blockchain technology in the agri-food system: A systematic bibliometric analysis and policy imperatives," 2021, Accessed: April 20, 2021. [Online]. Available: https://papers.ssrn.com/sol3/papers.cfm?abstract_id=3814912.
4   W. Liu, X. F. Shao, C. H. Wu, and P. Qiao, "A systematic literature review on applications of information and communication technologies and blockchain technologies for precision agriculture development," *J. Clean. Prod.*, vol. 298, 2021, doi:10.1016/j.jclepro.2021.126763.
5   M. A. Ferrag, L. Shu, X. Yang, A. Derhab, and L. Maglaras, "Security and privacy for green IoT-based agriculture: Review, Blockchain solutions, and Challenges," *IEEE Access*, vol. 8, pp. 32031–32053, 2020, doi:10.1109/ACCESS.2020.2973178.
6   M. Torky, and A. E. Hassanein, "Integrating blockchain and the internet of things in precision agriculture: Analysis, opportunities, and challenges," *Comput. Electron. Agric.*, vol. 178, p. 105476, 2020, doi:10.1016/j.compag.2020.105476.
7   H. Xiong, T. Dalhaus, P. Wang, and J. Huang, "Blockchain technology for agriculture: applications and rationale," *Frontiers in Blockchain*, vol. 3, 2020, doi:10.3389/fbloc.2020.00007.
8   S. Ahmed, M. E. Islam, M. T. Hosen, and M. H. Hasan, "BlockChain based fertilizer distribution system: Bangladesh perspective," *ACM Int. Conf. Proceeding Ser.*, 2020, doi:10.1145/3377049.3377116.
9   V. S. Yadav, A. R. Singh, R. D. Raut, and U. H. Govindarajan, "Blockchain technology adoption barriers in the Indian agricultural supply chain: an integrated approach," *Resour. Conserv. Recycl.*, vol. 161, p. 104877, 2020, doi:10.1016/j.resconrec.2020.104877.
10 Y. Chen, Y. Li, and C. Li, "Electronic agriculture, blockchain and digital agricultural democratization: Origin, theory and application," *J. Clean. Prod.*, vol. 268, 2020, doi:10.1016/j.jclepro.2020.122071.
11 D. Prashar, N. Jha, S. Jha, Y. Lee, and G. P. Joshi, "Blockchain-based traceability and visibility for agricultural products: A decentralizedway of ensuring food safety in India," *Sustain.*, vol. 12, no. 8, 2020, doi:10.3390/SU12083497.

12 L. Hang, I. Ullah, and D. H. Kim, "A secure fish farm platform based on blockchain for agriculture data integrity," *Comput. Electron. Agric.*, vol. 170, no. December 2019, p. 105251, 2020, doi:10.1016/j.compag.2020.105251.

13 X. Li, D. Wang, and M. Li, "Convenience analysis of sustainable E-agriculture based on blockchain technology," *J. Clean. Prod.*, vol. 271, p. 122503, 2020, doi:10.1016/j.jclepro.2020.122503.

14 K. Dinesh Kumar, D. S. Manoj Kumar, and R. Anandh, "Blockchain technology in food supply chain security," *Int. J. Sci. Technol. Res.*, vol. 9, no. 1, pp. 3446–3450, 2020.

15 L. Song, X. Wang, and N. Merveille, "Research on blockchain for sustainable e-agriculture," *2020 IEEE Technol. Eng. Manag. Conf. TEMSCON 2020*, vol. 7, 2020, doi:10.1109/TEMSCON47658.2020.9140121.

16 A. Vangala, A. K. Das, N. Kumar, and M. Alazab, "Smart secure sensing for IoT-based agriculture: Blockchain perspective," *IEEE Sens. J.*, vol. 1748, no. c, pp. 1–1, 2020, doi:10.1109/jsen.2020.3012294.

17 S. S. Kamble, A. Gunasekaran, and R. Sharma, "Modeling the blockchain enabled traceability in agriculture supply chain," *Int. J. Inf. Manage.*, vol. 52, November 2018, pp. 1–16, 2020, doi:10.1016/j.ijinfomgt.2019.05.023.

18 P. W. Khan, Y. C. Byun, and N. Park, "IoT-blockchain enabled optimized provenance system for food industry 4.0 using advanced deep learning," *Sensors (Switzerland)*, vol. 20, no. 10, pp. 1–24, 2020, doi:10.3390/s20102990.

19 M. Shyamala Devi, R. Suguna, A. S. Joshi, and R. A. Bagate, "Design of IoT blockchain based smart agriculture for enlightening safety and security," *Commun. Comput. Inf. Sci.*, vol. 985, pp. 7–19, 2019, doi:10.1007/978-981-13-8300-7_2.

20 A. Kamilaris, A. Fonts, and F. X. Prenafeta-Boldú, "The rise of blockchain technology in agriculture and food supply chains," *Trends in Food Science & Technology*, vol. 91, *arXiv*, pp. 1–33, 2019.

21 S. Umamaheswari, S. Sreeram, N. Kritika, and D. R. Jyothi Prasanth, "BIoT: Blockchain based IoT for agriculture," *Proc. 11th Int. Conf. Adv. Comput. ICoAC 2019*, pp. 324–327, 2019, doi:10.1109/ICoAC48765.2019.246860.

22 J. Hua, X. Wang, M. Kang, H. Wang, and F. Y. Wang, "Blockchain based provenance for agricultural products: a distributed platform with duplicated and shared bookkeeping," *IEEE Intell. Veh. Symp. Proc.*, vol. 2018, no. Iv, pp. 97–101, June 2018, doi:10.1109/IVS.2018.8500647.

23 L. Ge, C. Brewster, J. Spek, A. Smeenk, and J. Top, *Blockchain for Agriculture and Food*. Wageningen: Wageningen Economic Research, 2017.

24 Y. P. Lin et al., "Blockchain: The evolutionary next step for ICT e-agriculture," *Environ. - MDPI*, vol. 4, no. 3, pp. 1–13, 2017, doi:10.3390/environments4030050.

25 G. Leduc, S. Kubler, J-P. Georges, and U, "Innovative Blockchain-based farming marketplace and smart contract performance evaluation," *J of C. Production, Elsevier*, Accessed: April 20, 2021. [Online]. Available: https://www.sciencedirect.com/science/article/pii/S0959652621012749.

7 The COVID-19 Outbreak
Blockchain-Secured Transactions in Supply Chain Orchestration

Sushant Kumar, Vinita Tiwari, Basant Agarwal, and Jagannath Jayanti

CONTENTS

7.1 INTRODUCTION

The novel Coronavirus disease (2019-nCoV) has posed a serious challenge to the entire world. The serious concerns with COVID-19 is the pace at which it is communicated from the infected to those who come in proximity with the person [1, 2]. Perhaps, it is not an exaggeration to say that there is no sector which has not been impacted by this unprecedented global outbreak of a viral infection. Though the major concern is not the mortality rate of the virus but the way it gets transmitted from an affected person to others. The increase in the number of cases and deaths have caused worldwide quarantine and lockdown.

The scientific community across the globe is actively pursuing a cure, treatment protocol and more aggressively for a vaccine. Although several clinical trials reported to date have shown promising results, a more concrete and viable solution is much awaited. Even though different vaccines like SputnikV, Covaxin, Pfizer, Covishield, etc. are being administered across the world, certainty for lifting the lockdown in totality and going back to lifestyles, preceding the time of the pandemic is yet to be achieved. Care still needs to be taken in terms of socializing and interacting with others. Hence, businesses as well have to operate under strict guidelines maintain proper hygiene and sanitation and also following the social distancing norms.

The number of people being attended and entertained by various businesses is also restricted as result of the abovementioned reasons. It has left serious economic

DOI: 10.1201/9781003107507-7

consequences. The longer the time period for which the virus promulgates, the impact it would on the economy. This also raises concerns regarding financial sustainability [3]. Many nations across the globe have resorted to lockdown either in a partial or complete sense (India is one such country) to stop the spread of the disease. This unprecedented move has left many people jobless and with the manufacturing sector seriously hit, global economic crisis looms around the corner. Cinema halls, restaurants, shopping complexes etc. were shutdown. In such times online shopping, seemed to be a viable alternative. However, because of the rate at which the virus transmits from one person to other who are in vicinity of one another, the odds of the virus spreading from delivery agents to consumers and vice versa could not be neglected. As a result of this, even the e-commerce services were also suspended in a number of regions across countries. Due to hinderance caused by these bleak times, a system needs to put in effect in order to tackle this situation in present time and also a system or methodology for keeping the various ventures up and running so that people at the bare minimum have access to basic amenities. Whatever solution or alternative attained should also be sustainable enough that in case of any such unprecedented situation is faced in the future the world can still operate. Extensive research is conducted in this subject in order to develop methods to overcome the ominous problem the world is currently facing.

The basis of this work revolves around the fact that in the era of globalization where the supply chains are spread across the globe there needs to be a better technology aided system which could replace the conventional supply chain management techniques while retaining the individual interests of the individual stakeholders. With the advances in technology, there is now scope for extensive study, research, and exploration to look for potential solutions for an effective maintenance of the supply chain. Blockchain is being proposed as an alternative method for maintaining the supply chain which remains the key in global crisis. In this work a review of the key aspects of blockchain technology and its potential to improve the supply chain management is done [4]. With its distributed nature, multiple records keeping, and working without a centralized service provider, blockchain helps maintain secrecy, protects data, transparency, and allows trading among various stakeholders in a consortium. Apart from the medical challenges another major concern during the COVID-19 crisis is its detrimental effect on the global economy. An official statement from the chief of the International Monetary Fund (IMF), Kristalina Georgieva, on 28 March 2020 explains the magnitude of the impact COVID-19 has on the world economy. The main impediment for the economy, however, is the nature of the virus transmission more than the virus itself. With social distancing norms in place across the globe, travel restrictions and anxiety among the consumers could be stated as the potential consequences for the worst recession the world has ever seen in the last 100 years.

One of most salient parts of a business to maintain the continuity of production and the business itself is the supply chain. It is an integral part to ensuring the success and growth of a company. Supply chain is crucial as it performs many functions like boosting the customer service, reducing operational costs (include decreasing the purchasing cost and decreasing the production cost), improve the financial position of an enterprise (which include increasing the profits, decreasing the fixed assets, and increasing the cash flow) [5].

Since, supply chain is of such monumental importance to any business it goes without saying that it needs to well managed. Supply chain management deals with each step or process involved, beginning from the origin of service or the product and terminates with the delivery or consumption by the end user. Hence in a nutshell it's a "network of business entities" which participate in the delivery of products and/or services to the end user.

A conventional supply chain management for the manufacturing sector is shown in Figure 7.1. Each block represents a "business entity." The process begins with the procurement of raw material followed by manufacturing of the product using the raw material and then followed by the delivery (distribution), warehousing (bulk storage of the products), retailer and, finally. the end user.

The following is a brief overview of each of the major elements of the supply chain.

Raw Material: These are the basic building blocks a product and indirectly the supply chain. Raw materials refer to the basic components, unprocessed materials which are required for manufacturing of a product. Raw materials are not just the building blocks, they are also the starting points for any manufacturing process. They also are an investment for the organization which is creating the product. As buying raw materials for manufacturing depends on the purchasing power of an organization, which is a measure of the financial strength of an enterprise. Therefore, it is of the utmost importance that companies employ adequate and effective measures for managing their inventory of raw materials.

Manufacturer: The manufacturer could be anyone who creates a product. From large enterprises like Apple producing smartphones, laptops to a small-tailors who stich cloth material into sophisticated suits are all manufacturers of their products operating different parts of a gigantic commercial spectrum spread all across the planet. Every organization, every enterprise, every business that manufactures a product, needs to make sure that whatever it is that they are building, paramount importance needs to be given to the quality of the product keeping in mind costs of production and revenue generated. Being meticulous in the process ensures maximum retention of customers and a thriving business.

Distributor: In a volatile world, where industries keep changing, new technologies invented at a rapid pace, distributors play a pivotal role in the management of the supply chain. In recent times, rather than ordering items in bulk, customers prefer to

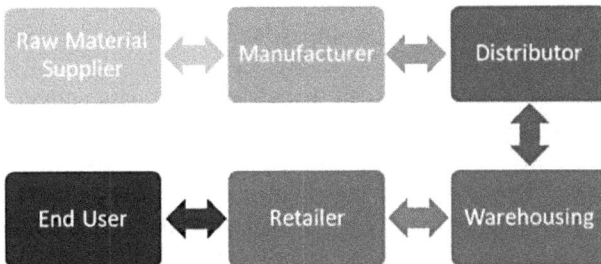

FIGURE 7.1   Conventional supply chain management for a manufacturing sector.

order in small quantities recurrently. In such a case continuous creation of products is a must. However, as importance it is to create a quality product, it is equally important to ensure that product reaches the buyer or the end user in proper conditions without causing damages to the product in the transit. The manufacturers need someone trustworthy to ensure proper delivery of their products. The distributors make sures of the delivery of the product. A consumer might wonder why pay more to product having distributor as a middleman. However, the fact is in return for less price they need to play the role of the distributor for themselves, the effort of which is much more than the extra money for buying the product. The distributors thus play a crucial role as they painstakingly maintain continuity of the supply chain.

Warehousing: Warehousing is an equally important aspect of the supply chain. With the advent of globalization and industrialization and huge growth in technology and machinery, the production has become humongous. The huge quantities being transported to various sites for manufacturing processed, and the commensurate amounts of products being produced is truly overwhelming. In this scenario, having everything in transit as and when needed is a very expensive and highly risky for organizations. The chances of the products getting damaged increase significantly in such a situation. Therefore, the products manufactured in such prodigious amounts need to be stacked and stored at appropriate locations based on a number of factors like transportation, amount of space available, etc. Warehouses come into the picture for this very purpose. In warehouse products are stored in bulk, and are sorted properly there which makes it easy to locate the items inside the warehouses.

Retailers: An entity that sells the product to the intended user. The retailer could be a business or person who sells the product created by the manufacturer. The retailer obtains the product from the distributor and in turn sells them to consumers. Retailers are the penultimate aspect of the supply chain from where the product reaches the place it was supposed to be.

End user: Last but not the least are the end users or consumers. They could be referred to as the most important or valuable part of the supply chain. It is for the consumers a manufacturer creates a product. The more the consumer is gratified with the product the more he/she buys and more the profit the manufacturer makes.

All the of the above together make up the supply chain work in consonance to ensure the smooth and harmonious flow and continuity of the supply chain.

This is a highly simplified model for the supply chain management without loss of generality. It is worth mentioning here that the size of the supply chain can be large or small and there can be individual supply chains within the entity too (For example, manufacturing can be a multi-step process where each step is dependent on the output from the previous step.)

The double-headed arrow between the adjacent business entities in the model indicates the communication link between the two. An important observation looking at the model is that the communication is possible only between the two adjacent "business entities." This leads to several problems. The major problem being the lack of transparency, information inadequacy to plan the next product development cycles and due to the globally distributed nature of the supply chains in several industries today with this model if there is a disruption in any one of the business entities due to a local issue the entire chain will be disrupted jeopardizing the entire industry. Had there been some information which is shared among the various business entities, there could be a possibility of making an alternate arrangement to make up for the losses and sustain until the regular services are resumed.

Supply chains are complex machines, with multiple stakeholders. According to the 2020 Supplier Information Study, 81% of supply chain leaders aren't completely confident in their supplier data, while 60% say it took them four days to update outdated supplier information and resulted in unhappy clients and financial loss [6]. The effects of this problem would increase manifold because of the global pandemic and the mandated restrictions. Some methodology to combat this issue is required. Though technologies like artificial intelligence (AI), machine learning, and blockchain can't fight the n-COVID they can help manage the economic systems well. The aim of this paper is to discuss the possibility of using block chain technologies to circumvent the challenges posed by COVID-19.

Blockchain has broken into the scene as a disruptive technology and could be one of the dominant tools of the industry in the near future. It provides a number of features which of prime importance in the present world like data privacy, transparency, decentralization, data immutability etc. These features improve processes involved in production and add to the currently supply chain by increasing its agility, resilience, and responsiveness. Therefore, enterprises should study and evaluate the comparison of blockchain oriented supply chains and traditional supply chains. The high value of global desirability index for supply chain with integrated blockchain compared to the traditional supply chains suggest that it justified to for use blockchain technology in supply chains to bring sustainability to them [7].

7.2 RELATED WORK

Fear of overburdened healthcare and the scarcity of resources due to global lockdowns are lifting on their shoulders the direct repercussions of supply chain breakdowns. According to the world economic forum, coronavirus 2019 exposed many weaknesses in the global supply chain management system [8]. The last five months of the pandemic have revealed the lack of connectivity and data exchange in the global supply chain [9]. As per the prognostication by JPMorgan Chase & Co., the Pandemic, COVID-19 has the potential to paralyze the global economy, with an estimated loss of more than 5.5 trillion US dollars in the next 18–24 months [10].

This breakdown in the collaboration required to track, authenticate, trace, finance, and clear medical goods, supplies, etc. via trade channels in a trusted, verifiable, and efficient manner [11]. COVID-19 has had a huge impact on the supply chain of globally connected networks. Pandemic highlights the need for an interconnected and interoperable supply chain in a world. Challenges are not only in terms of technology, but trade restrictions further intensify the supply chain disruptions causing a majority of suppliers having to halt production and several logistic partners to postpone the transport of goods. Many organizations have "One Up" and "One Down" visibility in the supply chain and because of that the food supply lacks true end-to-end visibility [12]. Access to trusted data in real time and as a result shortage of food or raised prices of essential goods can be seen in many areas. Due to lack of visibility in the supply chain, many retailers are storing the food rather than providing it to customers and then charging huge amounts of money for them. With the use of blockchain and Internet of Things (IoT), food companies and retailers can be connected with the government and the government can supervise the food supply chain and can have end-to-end visibility and trust [13].

Therefore, many countries are exploring blockchain technology for smooth delivery of services to their people. The UAE is one such country which is gradually transitioning to blockchain technology for smooth distribution of government services [14]. UAE pass is one such application developed by the UAE government. Blockchain technology is being deployed by China to combat the situation of pandemic, applications include charitable donation management, medical supplies providing finance to small and medium-sized companies. Centralization of essential ingredients for drugs in China makes it vulnerable to interruption, therefore China employed block chain technology in data analytics for healthcare during the time of pandemic [15]. Again, Sydney is adopting blockchain along with IoT and 5G which is going to contribute AU\$ 30 billion to their economy by 2029 [16].

Management of supply chains and controlling them becomes more difficult with their globalization. Blockchain aids in this process as it being distributed ledger technology ensures transparency, security, and traceability. It is proving to be credible in easing the difficulty to handle problems encountered in supply chain management. The increasing pressure to achieve sustainable goals provides even more motivation to study how blockchain can be employed to assist in the process of supply chain management [17].

Blockchain is becoming increasingly popular and has managed to gain the attention of a variety of industries. With a number of people having a positive attitude toward accepting this new technology, much detailed surveys and use cases can be explored to understand more as to how it can have a profound impact on the supply chain management [18].

A paraphernalia of benefits to show that blockchain can rather prove to be a benefactor in supply chain management have been identified. Some of them are improving sustainable chain management, reducing the cost of transactions in the supply chain, reduce illegal counterfeiting. However, since the technology is very much in its nascent stages, the researchers in the field of supply chain management need to investigate further into much more depth as to how blockchain can be used with respect to supply chain management [19].

Apart from blockchain, AI has also been introduced to help in the supply chain management. AI has been used to analyze data, reveal business patterns, and help in gaining much better insights into the supply chain [20].

With huge amounts of data being generated in present world, that too at a rate never seen before, research needs to be conducted regarding supply chain management. Data science techniques can be used to understand data. Methods to reveal patterns and trends in the data, to understand the various aspects of supply chain with much better insights could prove to be really helpful to develop more agile systems.

JH Tseng et al. [21] suggested the use of Gcoin blockchain. It was suggested as the base for flow of data related to drugs in order to generate transparent data regarding drug transactions. All the units involved in the drug supply chain would be able to simultaneously indulge in order to prevent spreading of counterfeit drugs with a goal of safeguarding public health.

Khalid et al. [22] developed a consensus algorithm that eliminates the necessity of a trustworthy centralized authority. They approached the agricultural supply chain with the Ethereum blockchain. Their algorithm performs optimally and facilitates transactions among all the members of the supply chain. The blockchain' immutable ledger is used to store and keep record of all the transactions.

The pharmaceutical supply chain is yet another sector where major breakthroughs and development using technology are being searched for. Supply chains powered by blockchain deliver plausible results in drug recall management and addressing prescription drug abuse. The capability to automate various processed in the considered supply chain can reduce costs by a substantial factor using smart contracts. Beyond just the pharmaceutical supply chain applications using blockchain can be put to use in areas dealing with medical supplies and devices and public health [23].

Modum.io, a start-up presented in the paper [24], used IoT devices with sensors using blockchain to facilitate accessibility of temperature records and also ensure data immutability in pharmaceutical supply chain, reducing operational costs at the same time. All the data regarding temperature and humidity to ensure quality control is added into the blockchain.

In the wine supply chain as well, the need of a safe and traceability system is inevitable. With increasing adulteration, use of preservatives and hazardous chemicals being used extensively and increasing counterfeiting the exigent need for such system is much more indispensable. In a blockchain-based supply chain every single transaction is recorded, and every relevant member can view those records in the form of blocks. On top of providing traceability, the system also provides safety, transparency, and security [25].

In the modern world, when there are multiple stakeholders, maintain a supply chain is rather challenging. The system currently used are having a number of flaws, like lack of reliable data. Blockchain has emerged has a potential alternative which can possibly cover these flaws of the traditional methods. Blockchain provides many features like security [26, 27], traceability, immutability, transparency and most importantly trust between the stakeholders of an enterprise [28].

As RFID tags can easily be cloned in any public space their credibility cannot be totally guaranteed in post supply chain. In the article [29], product ownership management system (POMS) is proposed. The motivation from the blockchain of bitcoin is used. Using the system customers have scope to reject fake products [30]. The system is developed using a blockchain-based decentralized application.

Blockchain along with other technologies helps to resolve the various challenges of the global supply chain. Blockchain as a shared ledger records and verifies transactions using strong encryption standards within a peer-to-peer network system. Transactions are decentralized and transparent, so every node in the network has its own copy of the information. This helps to eliminate any single point of failure and fraud.

7.3   PROPOSED ARCHITECTURE

An alternative model representing the blockchain-based supply chain management for the manufacturing sector example presented in Figure 7.1, and Figure 7.2. It can be seen that there is a common connection established between various "business entities" through blockchain technology. However, it has to be remembered that in

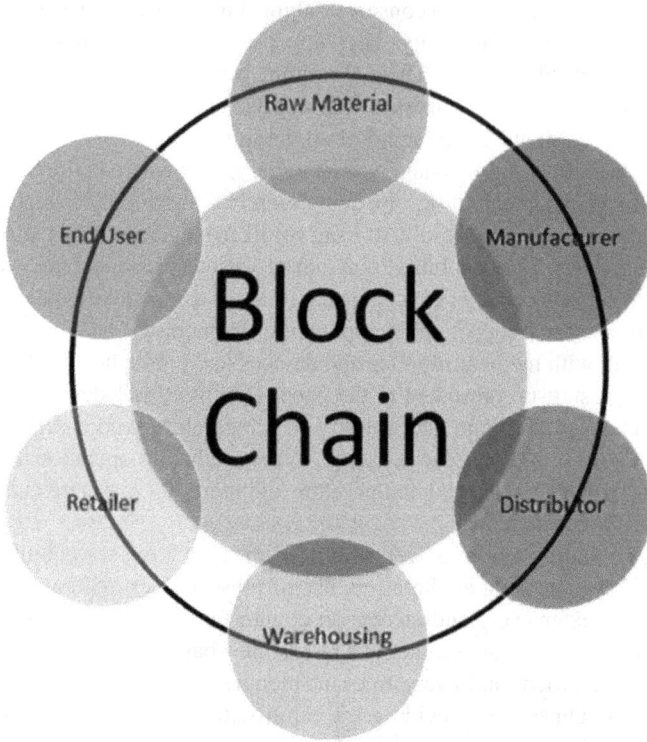

FIGURE 7.2   Blockchain-based supply chain management for the manufacturing sector.

this technology, there is no centralized server or a record keeping mechanism which has access to all the information which could possibly hurt the individual interests of the stakeholders. The information is shared in a decentralized way but is shared with each of the stakeholders on a selective basis.

Blockchain has two kinds of model, permissionless and permissioned. In the permissionless, there are a large number of participants, and their identity is anonymous; this type is used generally for mining cryptocurrencies. The other model is permissioned, where only a few participants are invited to form a group or a consortium for any business activity and their identity is authenticated beforehand. Every party knows every other party and they work in an amicable environment. However, the trust between each other for an equitable trading may not be prevalent. So, the blockchain framework comes into play. A blockchain framework can revolutionize global trade among different parties involved. Blockchain properties like immutability, transparency, security, and automation enable supply chain orchestration.

Before reaching the end consumer, the goods travel through a vast network of participants like gathering raw material, manufacturing goods, distribution center, government auditors, containers, ports, retailers and then, ultimately, it reaches the end user consumer, as illustrated in Figure 7.3. In this long chain of supplies, there

End-To-End Blockchain-Enabled Supply Chain

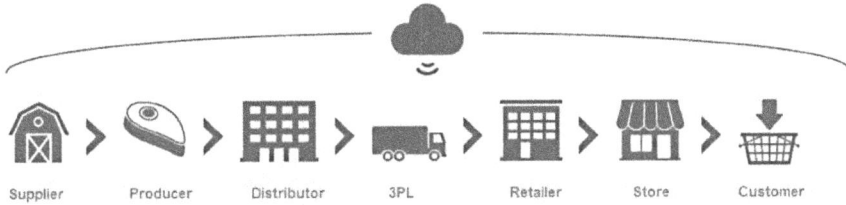

FIGURE 7.3 Vast supply chain of intermittent stoppages. (Source: ref. [5]).

are hordes of people attached and their intermittent intervention is imminent. Covid-19 precautions have inadvertently put the brakes on this chain. For the economy to revive and be on track, these obstacles need to be removed.

In Figure 7.3, there can be a few potential vulnerabilities like integration, traceability, auditability, immutability, and automation at all stoppages. A hyperledger framework envisages these in a viable way. By sharing goods' details over the framework, end users can automatically self-verify the goods' authenticity.

Hyperledger Fabric, a permissionless blockchain, uses distributed ledger technology (DLT). It contains smart contracts, which are codes with logic, also called chaincode, describes the system's application logic. The other key elements of Fabric are membership service provider (MSP), ledgers, and peers (nodes) consisting of endorsers, committers and orders [31].

All organizations involved in the consortium have multiple peers, which contain ledgers involving transactions. Smart contract contains chain code and events. MSP contains digital certificates for all peers of the organization. It authenticates them of right privileges for access. The Fabric client interacts with blockchain. The ledger stores all the chain of blocks that keep all immutable historical records. The committing peers maintain the ledger and also commit the transactions [31]. Endorsers are responsible for executing the chain code, giving output, and then agreeing whether the transaction is legitimate or not. It also holds smart contracts. Ordering nodes, on the other hand, order all transactions, approve the inclusion of transaction blocks into ledgers, and communicate with both committing peers and endorsing peers.

As seen in Figure 7.4, the consensus is achieved by following the transaction flow: endorse, order, validate. A client application invokes transactions. The endorsement policy specifies which endorsers have to sign it for policy to come in effect. Every user has its unique client application id from the MSP. It sends multiple endorsing peers, whichever are of relevance to them. All the endorsing peers, as determined by policy, will sign their output. Those then will execute the transaction but not commit it. Each node or peer will capture the read write sets (RW) and sign it. Now the transactions are encrypted. The RW sets are sent back to the application. Now the client application submits responses to orderers. In parallel, other applications are also submitting transactions to orderers. The orderers also have a provision of channels which

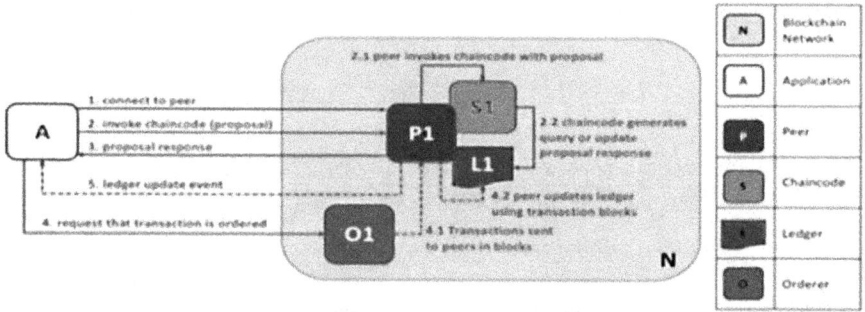

FIGURE 7.4   Hyperledger Fabric transactions. (Source: ref. [32]).

provide privacy among different ledgers. The orderers collect transactions into proposed blocks for distribution to committing peers.

Every committing peer validates against endorsement policy. These validated transactions are applied to the ledger and retained there. These committed and validated transactions will be added as a block in blockchain. In any event of failure, peers can emit events back to the application to notify it of the same.

Here, in Figure 7.5, it can be seen that a manufacturer who produces goods items, once he has all the details about it, logs into the client application, through MSP, identifying himself, also puts up prices of the goods and then other available organizations acting as peers verify the client signature and endorse the transactions. Then those transactions are ordered and sent to committing peers for validation. Finally, it updates in the ledger of the business activity. The ledger is sent to the manufacturer for shipping the goods item. The shipped item information is sent to the concerned peers through channels and whole activity status is being watched upon on the

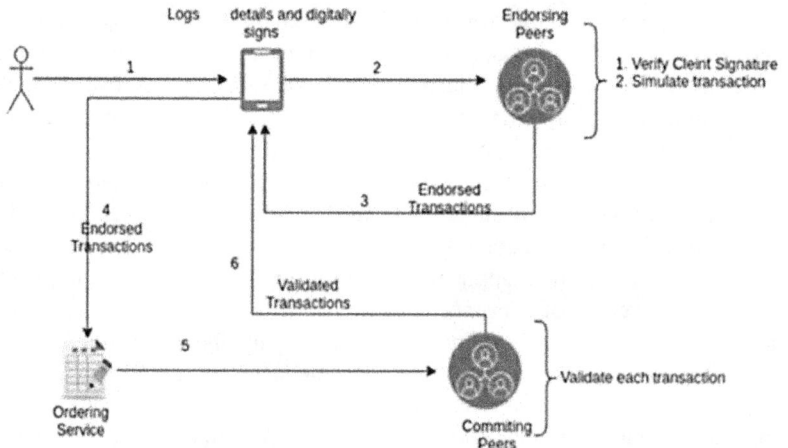

FIGURE 7.5   Smart contracts in action. (Source: ref. [33]).

blockchain. IoT helps to find goods' location at all times and is integrated inside blockchain. Any physical contact between stakeholders is precluded and an electronic digital interchange is set up.

The containers containing the shipments have IoT devices set up and installed, so as to monitor the location through GPS at all times. The location is trackable and is transparent to all the entities in the blockchain network. Also, the food retailers can monitor the temperature and humidity of specific degradable food items over the course of time. A real-time inventory and shipment is visible. After the shipment reaches the warehouse, it can be located easily with the help of IoT. Due to real-time location tracking, workers can also find the specific aisle for a particular item. IoT enables a seamless workflow with least manual intervention. The workforce which sets up the shipment has to maintain safety protocols like sanitizing all equipment for packaging, Personal Protective Equipment (PPE), masks, etc. Once it is shipped safely, the distributor has to follow the same pattern in taking goods back to his warehouse. Distributors can then connect with retailers on the blockchain network and make business deals work.

In Figure 7.6, all the stakeholders in the system – such as manufacturer, distributor, regulator, shipper, retailer, and consumer – are shown as connected to the Hyperledger Fabric network and all business activities are done through it. A manufacturer can sell his goods item by invoking a smart contract among other peers among the system. He will give details about the food items like unique ID, time, location, quality, weight, container number, etc., and at the same time the auditor/regulator can check and

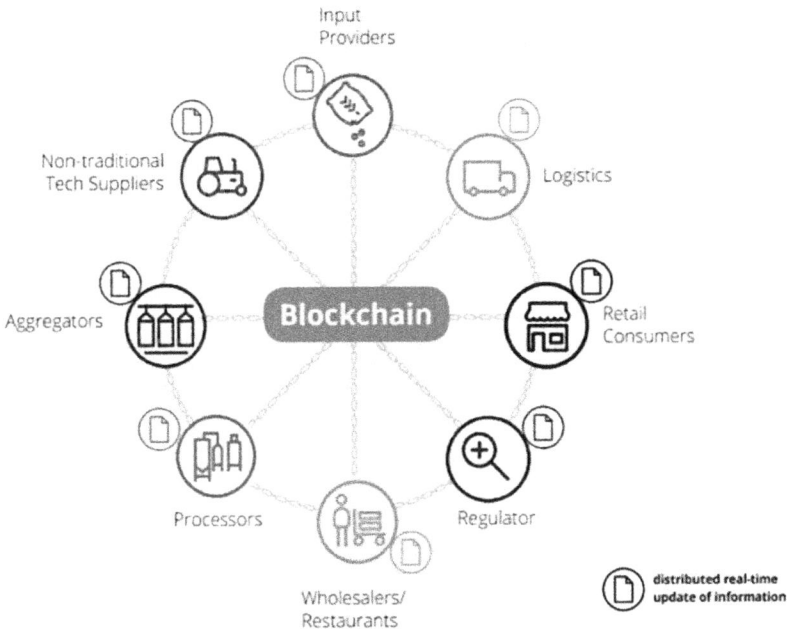

FIGURE 7.6   Blockchain network. (Source: ref. [34]).

verify if he is in ambit of government regulations. Once he is verified, other peers can check his smart contract endorsement and if willing to buy they can endorse him for an amount as asked by him. Also, a manufacturer can sell the same goods item to different peers or organizations at different prices through channels as mentioned before. This information is kept private between the manufacturer and a distributor. Everyone can run transactions between them and can fulfill their business demands. All the data is kept as transactions on blocks and maintained in a blockchain which is ensured in hyper ledger fabric itself. There is non-repudiation and digital signature at all times. Hence an ecosystem is set up inside and business can flourish.

In the COVID-19 scenario, all businesses are hampered with their productivity as social distancing is in place. There are minimum workers allowed and safety protocols are in place like masks and sanitizers. In these cases, the gathering of raw material, production, shipment, selling all are stuck at a very low pace. So, this Hyperledger Fabric network can be really effective because there is minimum manual intervention needed. There is no need for human gatherings, meetings, groups, contract papers, signing, etc. This will boost the business activities which are stuck at a snail's pace.

7.4 RESULT ANALYSIS

In the paper [35], a manufacturing supply chain is being discussed where a product's lifecycle is started by sharing and storing secure records of exchangeable data. As seen in Figure 7.7, each product will have its unique data like RFID, barcode, etc. So, a virtual identity (unique digital cryptographic identifier) will be set up as against its physical form. These details will be shared by all the beneficiaries. The beneficiaries can be certifiers (CA), manufacturer, distributor, shipper, consumer, retailer, etc.

Also, the beneficiaries will have their profile registered and their all details shared and stored in the organization's tab. Also, they are certified by a registered auditor.

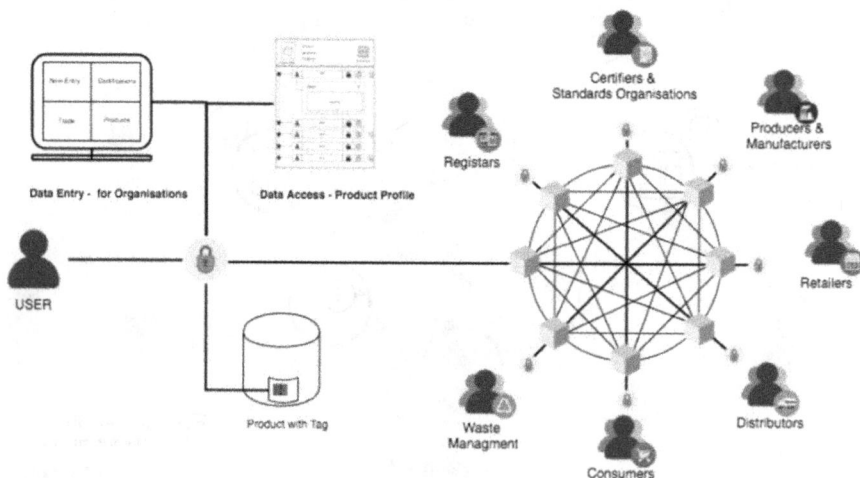

FIGURE 7.7 Blockchain-enabled supply chain. (Source: ref. [35]).

Any entity in the system, can view these details (if the profile is public); otherwise they can be anonymous too (still certification is mandatory) from a registered auditor able to make a deal. All beneficiaries have a public, private key pair. The public key helps them to communicate in the network while the private key digitally signs the product contract code. This software is run on a blockchain network such as Ethereum blockchain. The access of data in the network system depends on the authorization role to the beneficiaries.

7.5 POSSIBLE CHALLENGES IN SUPPLY CHAINS USING BLOCKCHAIN

The entire concept of blockchain is based on the idea of an immutable ledger which is distributed which has the potential to run businesses in a trustable and transparent way without the intervention of any central authority. Blockchain has shown great promise in adding great value to various industries, including the area under discussion which is the supply chain [36, 37]. As hyped as blockchain is, a more realistic perspective is required in order to assess various aspects related to it more objectively. Blockchain incurs high costs and high overheads in storage. It is primarily used for storing data related transactional data where trust is a lacking factor. Blockchain is a feasible option only when the costs realized could be outweighed by the benefits it has to offer.

The very first challenge could be high computational overhead when dealing with blockchain. Even though blockchain applications used in supply chain management are much more efficient than in cryptocurrencies, still high computing capabilities are required, for instance, for reading from the blockchain. These operations pertaining to applications using blockchain could be slower than those with traditional databases.

Since blockchain is decentralized, which is a significant advantage of it, managing it could be tiresome. Maintaining a decentralized application could prove to be far more difficult than managing centralized applications. Blockchain at the moment is not highly popular with enterprises as such.

Blockchain keeps a record of every transaction that takes place in the distributed ledger, and this can start compounding when used by multiple parties. The very algorithms that are used for encryption, to generate trust are the very algorithms which add data overhead. Handling data complexity is yet another challenge faced by applications using blockchain. Organizations because of the huge amounts of data they work could potentially end up managing much larger data structures than before. The early blockchain operated on business logic level of the applications which had to push both data and metadata to the blockchain. Enterprises therefore need to tackle this caveat when dealing with blockchain applications. They can reduce the risk of encountering this issue by striking a balance between data stored in traditional databases versus the data stored in the decentralized ledger [38].

One of the most distinctive features that sets blockchain apart from other data processing technologies is data immutability; however, data in supply chains is entered by people. Hence there is dependency on people to add quality data, because otherwise it could be very tedious to fix incorrect data entered into a blockchain application. The very essence of supply chains is the integration of different aspects of a business. To make these integrations, supply chains rely on various tools and adding a blockchain platform to all of them can prove to be a tricky task. It is because not every system is currently compatible with blockchain, and everyone is yet to accept to integrate blockchain into systems [39].

Blockchain as secured as it could be also faces a couple of security problems. One such problem is "the majority attack." If any entity at any point has more than 51% of the computing power, they have the ability to assume control of blockchain. Another issue is fork problems. These problems are related to the version of the node when software is due an upgrade. Forks are of two types: soft and hard [40].

Another major issue with blockchain is the problem of scalability. The traditional transactions systems are accepted widespread because of their speed of processing transactions. On the other hand, Bitcoin and Ethereum, the two largest networks running on blockchain are lacking by a huge margin in this aspect. There is also not much clarity about blockchain as such, which acts as a significant barrier for it to be adopted by masses [41].

7.6 CONCLUSION

These are unprecedented time. The global pandemic due to the novel coronavirus has made us realize that no matter how advanced the world, with increasing technology specifically in areas like AI, machine learning, robotics, etc. the world can never be fully prepared or equipped for such situation. The pandemic revealed many flaws in number of sectors and raised a number of questions modern science and its ability. The entire world was forced to temporarily go into a lockdown until the situation improved. Because of the lockdown and quarantine, the lives of millions of people were affected. People lost their livelihood, jobs, got entirely separated from others, particularly their loved ones living in different cities. A solitude that no one chose yet everyone was compelled to be in it. A huge number of people got affected by this mentally as well. Getting healthcare for ailments other than suffering from corona virus also became increasingly difficult as the masses were skeptical to step out of their houses and even more frightened to enter the hospitals because of the number of COVID patients admitted there. Apart from healthcare, getting the basic amenities also seemed to be a daunting task thanks to the lockdown and restrictions. From a world filled with busy roads, heavy traffic, the new normal was pointing in a direction where even the busiest of roads would be empty.

In such times, the onus of keeping the world running fell on a number of industries. Some of them are healthcare, tech industry, and many businesses producing everyday goods. Healthcare experts are actively looking for vaccines and other methods to both prevent the spread and cure the patients suffering with coronavirus. Technology, however, as advanced as it can be, cannot fight the novel coronavirus. Fortunately, it played a very crucial to help the functioning of the world until the vaccine was released. It because of technology, norms like work from home came into reality. Also, technology aided businesses to keep operating, and as the motive of this paper is, blockchain proves to be a very viable option for supply chain management both for the present and the future. When the world is reeling under severe novel corona, all that is needed is safety and smooth flow of business workplace. A consortium is created among different organizations/stakeholders interested in business continuity. At this time, there is a shortage of manpower and there are efforts to be precarious in this situation. Amid this, fulfilling all goals of safety of people and moving ahead with materialistic demands in society, blockchain can be a savior. In this work a model is proposed where there is least manual intervention and business can grow rapidly too. Any shipment of goods can be carried out from a manufacturer to a consumer via amiable work environment and trust among stakeholders. This increases the efficiency and growth of the whole ecosystem. The end-to-end transaction traceability and transparency will enhance trust throughout the supply chain system. This process automation will save costs and time. In a nutshell, the goal of this paper was to provide an architecture which could ensure that the entire process of a business, from a manufacturing of a product to the end user consuming it, could be achieved with minimal human intervention. This enhances efficiency and trust across the supply chain, saving huge amounts of expenditure and time, and also helping businesses to expand and grow.

REFERENCES

[1] Sarfraz, Z., Sarfraz, A., Iftikar, H. M., & Akhund, R. (2021). Is COVID-19 pushing us to the fifth industrial revolution (Society 5.0)?. *Pakistan Journal of Medical Sciences*, 37(2), 591.

[2] Bhatnagar, V., Poonia, R. C., Nagar, P., Kumar, S., Singh, V., Raja, L., & Dass, P. (2020). Descriptive analysis of COVID-19 patients in the context of India. *Journal of Interdisciplinary Mathematics*, doi:10.1080/09720502.2020.1761635.

[3] Gupta, M., Abdelmaksoud, A., Jafferany, M., Lotti, T., Sadoughifar, R., & Goldust, M. (2020). COVID-19 and economy. Dermatologic therapy.

[4] Baudet, C., Medina, M. J., & Lebraty, J. F. Blockchain and Supply Chain in Turbulent Context *IEEE Access*, 4(14), 2292–2303.

[5] Kohl, H., Henke, M., & Daus, D. (2021). 02 The importance of supply chain management for sustainability in global value chains. Sustainability in Global Value Chains: Measures, Ethics and Best Practices for Responsible Businesses, 3.

[6] Jairoun, A. A., Al-Hemyari, S. S., Abdulla, N. M., El-Dahiyat, F., Jairoun, M., Al-Tamimi, S. K., & Babar, Z. (2021). Online medication purchasing during the Covid-19 pandemic: a pilot study from the United Arab Emirates. *Journal of Pharm Policy and Practice*, 14, 38.

[7] Mukherjee, A. A., Singh, R. K., Mishra, R., & Bag, S. (2021). Application of blockchain technology for sustainability development in agricultural supply chain: Justification framework. *Operations Management Research*, 1079, 1–16.

[8] Ivanov, D. (2021). Supply chain viability and the COVID-19 pandemic: A conceptual and formal generalisation of four major adaptation strategies. *International Journal of Production Research*, 59(12), 1–18.

[9] Alsamhi, S. H., Lee, B., Guizani, M., Kumar, N., Qiao, Y., & Liu, X. (2021). Blockchain for Decentralized Multi-Drone to Combat COVID-19. arXiv preprint arXiv:2102.00969.

[10] Ceccato, I., Palumbo, R., Di Crosta, A., Marchetti, D., La Malva, P., Maiella, R., ... & Di Domenico, A. (2021). "What's next?" Individual differences in expected repercussions of the COVID-19 pandemic. *Personality and Individual Differences*, 174, 110674.

[11] Liang, S., Liu, H., Gu, Y., Guo, X., Li, H., Li, L., ... & Tao, L. (2021). Fast automated detection of COVID-19 from medical images using convolutional neural networks. *Communications Biology*, 4(1), 1–13.

[12] Karmaker, C. L., Ahmed, T., Ahmed, S., Ali, S. M., Moktadir, M. A., & Kabir, G. (2021). Improving supply chain sustainability in the context of COVID-19 pandemic in an emerging economy: Exploring drivers using an integrated model. *Sustainable Production and Consumption*, 26, 411–427.

[13] Singh, S., Kumar, R., Panchal, R., & Tiwari, M. K. (2021). Impact of COVID-19 on logistics systems and disruptions in food supply chain. *International Journal of Production Research*, 59(7), 1993–2008.

[14] Papadaki, M., & Karamitsos, I. (2021). Blockchain technology in the Middle East and North Africa region. *Information Technology for Development*, 27(3), 1–18.

[15] Hasan, M. R., Deng, S., Sultana, N., & Hossain, M. Z. (2021). The applicability of blockchain technology in healthcare contexts to contain COVID-19 challenges. Library Hi Tech.

[16] Sartipi, F. (2021). Publicizing construction firms by cryptocurrency. *Journal of Construction Materials*, 2, 3.

[17] Saberi, S., Kouhizadeh, M., Sarkis, J., & Shen, L. (2019). Blockchain technology and its relationships to sustainable supply chain management. *International Journal of Production Research*, 57(7), 2117–2135.

[18] Hackius, N., & Petersen, M. (2017). *Blockchain in logistics and supply chain: Trick or treat?. In Digitalization in Supply Chain Management and Logistics: Smart and Digital Solutions for an Industry 4.0 Environment. Proceedings of the Hamburg International Conference of Logistics (HICL)*, Vol. 23 (pp. 3–18). Berlin: epubli GmbH.

[19] Cole, R., Stevenson, M., & Aitken, J. (2019). Blockchain technology: implications for operations and supply chain management. *Supply Chain Management: An International Journal*, 24(4), 469–483. doi:10.1108/SCM-09-2018-0309.

[20] Min, H. (2010). Artificial intelligence in supply chain management: Theory and applications. *International Journal of Logistics: Research and Applications*, 13(1), 13–39.

[21] Tseng, J. H., Liao, Y. C., Chong, B., & Liao, S. W. (2018). Governance on the drug supply chain via gcoin blockchain. *International Journal of Environmental Research and Public Health*, 15(6), 1055.

[22] Salah, K., Nizamuddin, N., Jayaraman, R., & Omar, M. (2019). Blockchain-based soybean traceability in agricultural supply chain. *IEEE Access*, 7, 73295–73305.

[23] Clauson, K. A., Breeden, E. A., Davidson, C., & Mackey, T. K. (2018). Leveraging blockchain technology to enhance supply chain management in healthcare: An exploration of challenges and opportunities in the health supply chain. *Blockchain in Healthcare Today*, 1(3), 1–12.

[24] Bocek, T., Rodrigues, B. B., Strasser, T., & Stiller, B. (2017, May). *Blockchains everywhere-a use-case of blockchains in the pharma supply-chain. In 2017 IFIP/IEEE symposium on integrated network and service management (IM)* (pp. 772–777). IEEE.

[25] Biswas, K., Muthukkumarasamy, V., & Tan, W. L. (2017). *Blockchain based wine supply chain traceability system. In Future Technologies Conference (FTC) 2017* (pp. 56–62). The Science and Information Organization.

[26] Jha, S., Kumar, R., Chatterjee, J. M., & Khari, M. (2019). Collaborative handshaking approaches between internet of computing and internet of things towards a smart world: A review from 2009–2017. *Telecommunication Systems*, 70(4), 617–634.

[27] Khari, M., Garg, A. K., Gandomi, A. H., Gupta, R., Patan, R., & Balusamy, B. (2019). Securing data in Internet of Things (IoT) using cryptography and steganography techniques. *IEEE Transactions on Systems, Man, and Cybernetics: Systems*, 50(1), 73–80.

[28] Jabbar, S., Lloyd, H., Hammoudeh, M., Adebisi, B., & Raza, U. (2020). Blockchain-enabled supply chain: analysis, challenges, and future directions. *Multimedia Systems*, 27, 787–806. doi:10.1007/s00530-020-00687-0

[29] Toyoda, K., Mathiopoulos, P. T., Sasase, I., & Ohtsuki, T. (2017). A novel blockchain-based product ownership management system (POMS) for anti-counterfeits in the post supply chain. *IEEE Access*, 5, 17465–17477.

[30] Chatterjee, J. M., Ghatak, S., Kumar, R., & Khari, M. (2018). BitCoin exclusively informational money: A valuable review from 2010 to 2017. *Quality & Quantity*, 52(5), 2037–2054.

[31] Stefanovic, N. (2021). Blockchain for Supply Chain Management: Opportunities, Technologies, and Applications. In *Encyclopedia of Organizational Knowledge, Administration, and Technology* (pp. 2472–2487). IGI Global.

[32] Kernahan, A., Bernskov, U., & Beck, R. (2021, January). *Blockchain out of the Box–Where is the Blockchain in Blockchain-as-a-Service?* In *Proceedings of the 54th Hawaii international conference on system sciences* (p. 4281).

[33] Alves, L., Carvalhido, T., Cruz, E. F., & da Cruz, A. M. R. (2021). *Using blockchain to trace PDO/PGI/TSG products.* In *Proceedings of the 23rd International Conference on Enterprise Information Systems (ICEIS 2021)*, vol. 2, pp. 368–376.

[34] Balamurugan, S., Ayyasamy, A., & Joseph, K. S. (2021). IoT-Blockchain driven traceability techniques for improved safety measures in food supply chain. *International Journal of Information Technology*, 1–12. doi:10.1007/s41870-020-00581-y

[35] Bhat, M., Qadri, M., Beg, N. U., Kundroo, M., Ahanger, N., & Agarwal, B. (2020). Sentiment analysis of social media response on the Covid19 outbreak. *Brain, Behavior, and Immunity*, 87, 136–137. doi:10.1016/j.bbi.2020.05.006.

[36] Abeyratne, S.A., & Monfared, R.P., 2016. Blockchain ready manufacturing supply chain using distributed ledger. *International Journal of Research in Engineering and Technology*, 05(09), pp. 1–10.

[37] Kumar, A., Liu, R., & Shan, Z. (2020). Is blockchain a silver bullet for supply chain management? *Technical Challenges and Research Opportunities. Decision Sciences*, 51(1), 8–37.

[38] Hobbs, J. E. (2021). Food supply chain resilience and the COVID-19 pandemic: What have we learned?. *Canadian Journal of Agricultural Economics/Revue canadienne d'agroeconomie*. doi:10.1111/cjag.12279

[39] Rana, N. P., Dwivedi, Y. K., & Hughes, D. L. (2021). Analysis of challenges for blockchain adoption within the Indian public sector: an interpretive structural modelling approach. Information Technology & People.

[40] Lin, I. C., & Liao, T. C. (2017). A survey of blockchain security issues and challenges. *International Journal Network Security*, 19(5), 653–659.

[41] Yaqoob, I., Salah, K., Jayaraman, R., & Al-Hammadi, Y. (2021). Blockchain for healthcare data management: Opportunities, challenges, and future recommendations. *Neural Computing and Applications*, 2260, 1–16.

8 Blockchain Technology in the Energy Sector

A Systematic Review of Challenges and Opportunities

M. Arvindhan, M. Thirunavukarasan, and A. Daniel

CONTENTS

8.1 INTRODUCTION TO BLOCKCHAIN TECHNOLOGY

Blockchains come in numerous structures and for the most part share four principal highlights: decentralized information stockpiling (a public record of exchange records), encryption, permanence, and an agreement calculation. As a specific kind of decentralized record innovation, blockchains store encoded information across shared networks, connecting together successive "blocks" of data into "chains." The data accessible to all organization members is a common record of all data exchanges on the blockchain. The agreement calculations guarantee that data is predictable and unchanging across this decentralized organization and dissuade individual clients from adding to record data without approval from the organization. Moreover, because of blockchain's design, earlier data on the chain can't be altered or eliminated, as doing so would compromise the respectability of the decentralized record [1].

DOI: 10.1201/9781003107507-8 **107**

8.2　OVERVIEW OF BLOCKCHAIN TECHNOLOGY

A default plot for blockchains goes like this: from passive to active. Blockchains can in any case be open to the public, as well as providing a clearly distinguishable record. A large number of digital currencies are used as free, decentralized blockchains (see Figure 8.1). Entirely private blockchains are provided to the individual member companies and used for anything like board monitoring or records. Private/public blockchain consortiums (or half-breed blockchains) can facilitate more efficient transactions by entities such as gatherings of financial institutions or healthcare providers, but they can also distribute data among members of these networks. A form of blockchain, in turn, depends on motivation and plans. Public and private blockchains concentrate on large groups; hence, they are cheaper and more available for individuals [2].

There is great variation in the way blockchains operate, depending on their design schemas, in terms of consensus algorithm, data accessibility, and immutability. Additionally, certain blockchain platforms build upon already existing blockchains (e.g. platforms like Ethereum) (i.e. the application level).

Other types of distributed ledger technology (DLT) include Hashgraph, DAGs, and tangles. Whereas ordinary DLTs rely on computational gossip to find agreement, a gossip-based D-Lite enjoys consensus and is organized like interweaving chains. Non-directed acyclic graphs permit parallel transactions and subtrees.

8.2.1　APPLICATION OF BLOCKCHAIN TECHNOLOGY

For a range of applications, the board can and has used data, ranging from information technology to medical to oil. There have been a number of blockchains during the three years' prior to 2017, and these are now broken down into three basic classifications: public, proprietary, and consortium.

As required (decentralized applications): These are at various stages of growth, and the most common ones are all involved in gaming or trading, as well as financing and payments. A cryptographic type of money can be implemented as a coin (crypto-based charge), or as utility (digital cash which runs on blockchains) (tokens that address an advanced security). (See Figure 8.2.) At this time there are three primary blockchain

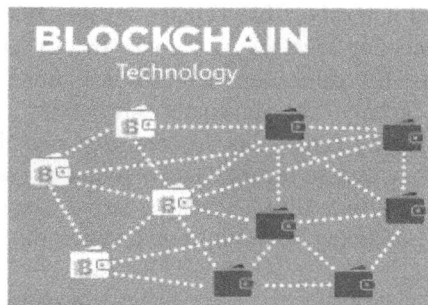

FIGURE 8.1　Blockchain technology.

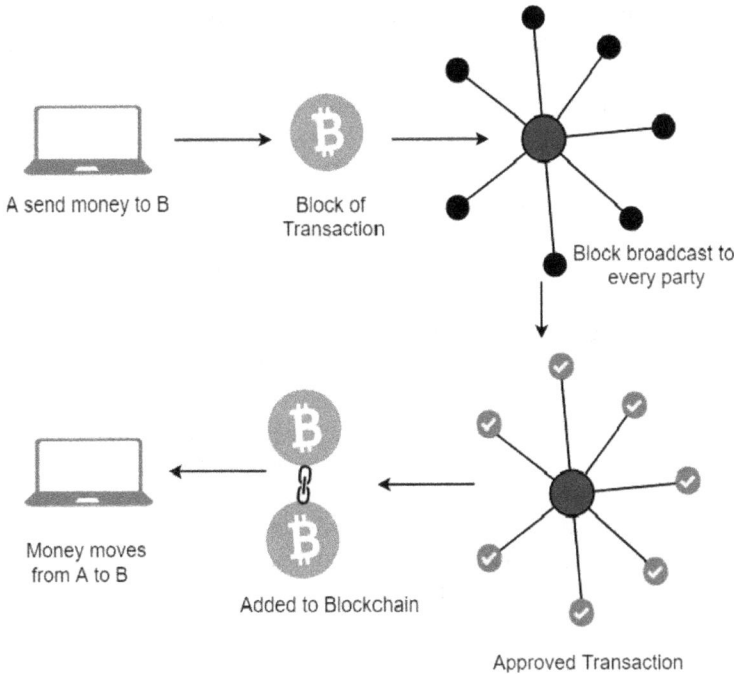

FIGURE 8.2 Transactions in blockchain technology.

variants: public blockchains, private blockchains, and custom blockchains; of the latter two, custom blockchains raise the most significant legal issues. As "clients" have pointed out, people can use the public blockchain as a speculative endeavor rather than in itself. Significant conversations about social networks include the Ripple (the Ripple conference) and block.one (blockchain.io) (the EOS public chain) [3].

8.3 ECOLOGY OF BLOCKCHAIN

According to the dictionary, the term "blockchain" may mean a record in which transactions in bitcoin or another digital currency are replicated and unchangeable. Although its locations are associated with high money, it is currently being lauded as a web with applications that span everything, for value rather than money. Some advantages of blockchain innovation are twofold: there is no single ledger that needs to be followed, and there is no single place of failure. Blockchain is increasing in use, and its impacts are more profound, and that's a guarantee of its future. So, overall, what are the consequences of this new discovery [4]?

The latter few months of 2014 also featured newsworthy instances of "digital money mining" being correlated with unintended consequences. Mining is the process of finding blocks, checking whether transactions are valid, and then broadcasting them in the blockchain (and is additionally the method by which new digital money is delivered). Everyone with access to the web has the right to engage in mining (Investopedia). To extract valuable ores, miners must use a tremendous amount of energy. It has been estimated that bitcoin has the same impact on the environment as 2.35 million cars. A new block is created and another process using energy is turned on each time the event occurs. While attempting to save resources, inefficient energy facilities are still found in China. As far as moving to a mining strategy is concerned, there is fuel source (and other necessities) variety, but in relation to electricity, we are proceeding from one side to the other [5, 6].

One organization, for instance, is building up an advanced cash and blockchain (Chia), which proposes to "ranch" instead of "mine" – the thought being that unused capacity on hard drives is used as opposed to handling power. GENERcoin is an "item supported computerized cash consolidating environmentally friendly power and digital forms of money." The cash, in this example, is upheld by "environmentally friendly power" energy. It may very well be reclaimed for the energy backing it, or exchanged a lot; likewise with some other digital money. EverGreenCoin is a cryptographic money used to raise assets for natural green undertakings – for example, renewables and water preservation [7, 8].

Blockchain has even been proclaimed as a method by which to ensure the climate; on the grounds that blockchain is available to numerous clients, information held is straightforward, can't be altered and once added can't be eliminated – it takes the approval of different PCs before another record can be added. This decreases exchange costs and go-between, as well as expanding productivity. Blockchain likewise takes into consideration distributed association. These things can be utilized to help the climate [9, 10].

8.3.1 FUTURE-THINKERS HAVE ASSEMBLED THE ACCOMPANYING MODELS

1. Decentralization will improve the ability to pursue ecological and international treaties.
2. Donations to good causes should be apportioned with accuracy and justice.
3. Products can be traced all the way back to their origin. This will help lower carbon footprints, improve environmental responsibility, and cut down on the number of questionable procedures.
4. It's a good idea to make recycling plans – for example, if you have rewards in place for participants.
5. In contrast to centralized distribution, it is conceivable to provide a point-to-point system with confined energy flow.
6. The blockchain can be used to track goods in order to assess carbon costs.

There is a growing amount of renewable generation, including wind and solar PV, that energy systems must account for. Many creative renewable energy sources have emerged recently due to privatization, as well as unbundling of the energy market and favorable financial policies. Environmental impacts of emerging energy sources have been widely studied in order to distinguish renewable and fossil fuel-based energy alternatives. It is difficult to integrate variable power from wind or solar farms into the overall power system. It is critical that those physical assets are dispersed in order to rethink our system of energy management. Possible models for adapting to the future renewable energy include a "honeycomb" of microgrids, some of which could combine wind, water, solar, and microtransactions. Blockchain technology is acknowledged to carry promise for legitimate and traceable transactions. Blockchain needs elastic tools, data storage and a development environment for developing applications, and on top of that, also requires elastic computing, as it empowers fast computing in short periods of time (preventive maintenance, energy management optimization, demand response applications). According to the writers, the conventional supply chain uses a lot of resources and creates tremendous emissions, which results in significant harm to the environment. A green supply chain emphasizes the greening of the entire process from beginning to end, and not just the products but also everything in the center, including design, manufacturing, distribution, recycling, and waste management. Therefore, businesses have chosen to highlight the low-carbon economy, integrate it into the supply chain, explain the future priorities of their enterprises, and enhance output in this low-carbon context [11, 12].

8.4   MULTI-NETWORK EMBEDDING AND GREEN SUPPLY CHAIN PERFORMANCE

Over the past decade, the growth of the green supply chain has gone from a single chain to a network of interconnected systems of various chains, made up of multiple stakeholders in the upstream and numerous consumers in the downstream. The multiscalar bed network has multiple combinations of node density and diversity of supply, and the non-chain performance, particularly in relation to embedding of heterogeneity of knowledge and resources, can offer businesses a number of different kinds of advantages. It promotes and maintains heterogeneous information and technical systems while preventing the "choking impact" of dense knowledge provision. More network awareness and activities within a single network could result in a lower innovation success rate compared to double or single networks [13, 14].

A blockchain is a decentralized and distributed database system that does not use a single point of failure or third-party trust. Smart contracts can be executed in a P2P network using blockchains (see Figure 8.3). Simultaneous updates allow for more users to be active, which in turn contributes to many variations of the ledger [15, 16]. Instead of handling the ledger through a single core of confidence, members reach an agreement based on a copy of the documents. We're still trying to decide exactly how consensus is achieved, but it could vary in different ways for various domains. Linking transactions to each subsequent transaction to the previous one allows for comprehensive and verifiable record keeping with blockchain. Any consumer on every network can do checks to see if transactions are right and trustworthy, and provide evidence. Blockchain has begun to realize its potential in multiple areas including startups, trials, testing, and studies in different industries. According to a survey commissioned by the German Federal Agency for Energy and Security, 20% of energy decision-makers agree that blockchain is a game-changer for energy suppliers. This is based on the opinions of 70 executives employed in the energy

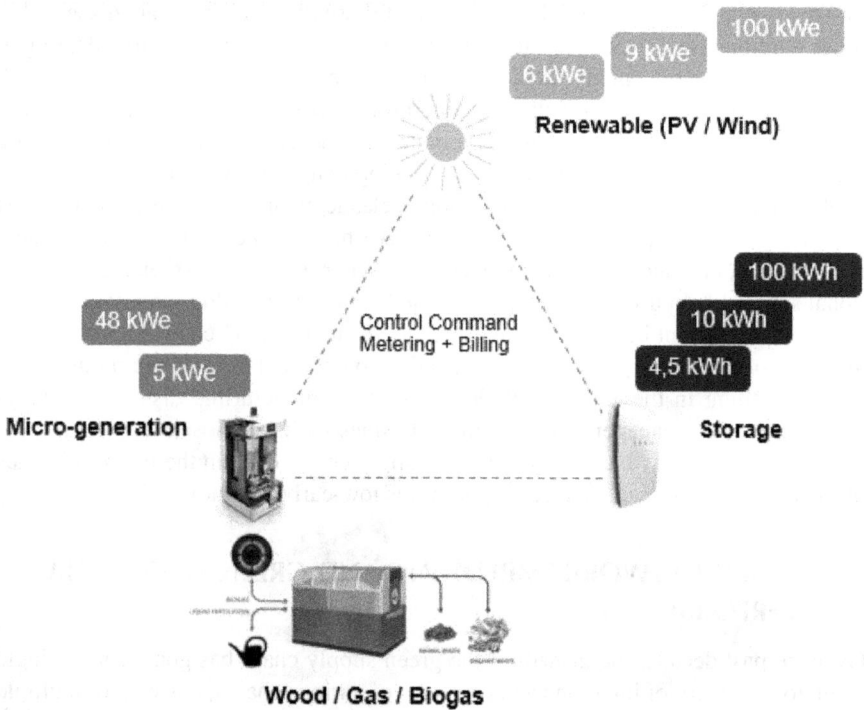

FIGURE 8.3 Green supply chain performance enabled with renewable energy.

industry, as well as energy providers, aggregators, utility, transmission, and generation suppliers [17].

Around half of participants have already started or are currently preparing efforts to incorporate blockchain or cryptocurrencies into their organizations or ventures. Many energy utilities are keenly interested in using distributed ledger applications (DLT) as a way to reduce carbon footprints and promote sustainability. Further, according to Deloitte and PwC senior business and technical consultancy reports, blockchains have the opportunity to disrupt several different industries due to the evolving cryptocurrencies as digital assets that can be exchanged across borders. This early research has raised hopes that blockchain could have the ability to address issues for the energy industry [18]. Three requirements can be identified for future energy systems: decarbonization, decentralization, and personalization. To date, however, the energy and electricity markets have proven unresponsive to this intent. As a result, a few players have basically been removed from the market, and financial incentives have failed to convince the active customer population to participate (see Figure 8.4). By building early transactional blockchain-based networks, early adopters are starting to put the fundamentals in place for peer-to-peer energy trading. They are working on integrating the local energy ecosystem and the Internet of Things (IoT) to further realize their concept of a smart grid. There is growing concern among the energy firms about increasing energy costs and declining revenue streams;

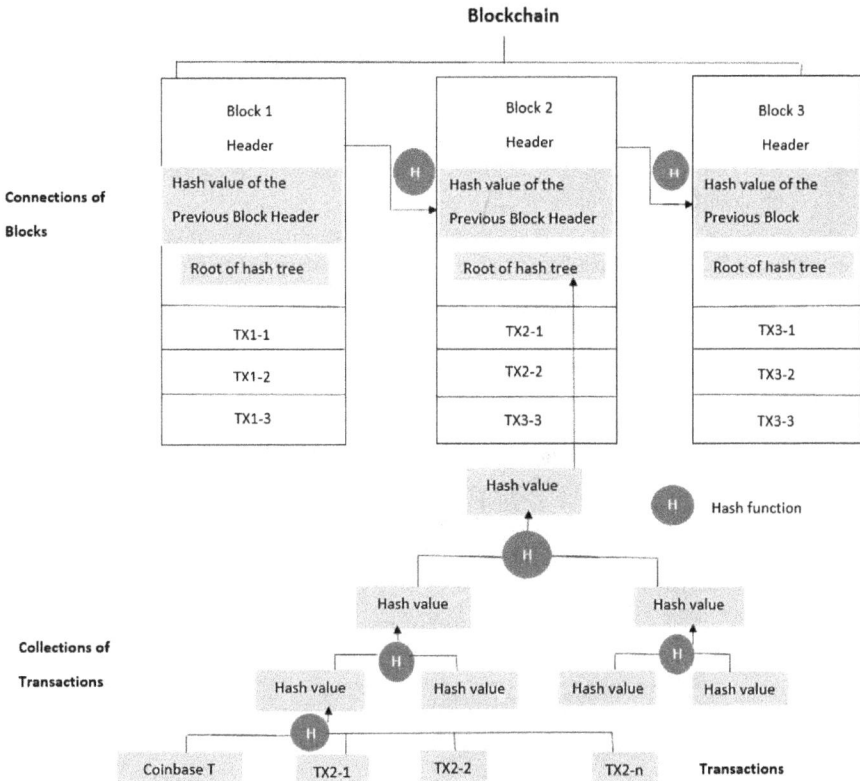

FIGURE 8.4 Transaction hash for block and header value in blocks.

according to PwC, the regulatory authorities' demand for increased accountability goes hand in hand with a corresponding increase in requirements for the creativity of the utility researchers. Any saving and efficiency in the operation of energy systems could be offset by possible safety issues, so it is safer to look into them all than to presume they do not exist [19, 20].

8.5 BLOCKCHAIN'S POTENTIAL IMPACT ON ENERGY COMPANY OPERATIONS

Blockchain technology is expected to have an impact on a range of market processes and activities related to the energy sector.

Given existing literature, below are the possible implementations and the characteristics of business models that could be impacted.

Using blockchains, smart contracts, and smart metering, customers and distributed generators can get real-time accounting. Energy micropayments, prepaid meters, and pay-on-use solutions can be useful for utilities.

Consumer energy profiles, priorities, and concerns about the environment, however, can significantly affect the way marketing is conducted. Blockchains in concert with AI techniques such as machine learning would be able to recognize user patterns, allowing for more and more personalized energy solutions.

Distributed trading platforms could challenge market management, commodity trading, and risk management operations. The hypothesis is that blockchains are used for carbon credits trading as well [21].

Decentralization and microgrids may benefit from the automation of blockchain. Re-localized P2P energy trading or on a distributed platform would allow substantial increases in local energy output and usage, impacting revenues and tariffs.

Blockchains can be used for smart device communication, data transmission, or both, including smart meters, sophisticated sensors, network monitoring systems, and facilities to regulate electricity all being part of the smart grid. Besides making information more secure, grid implementations can also provide numerous other benefits, such as information standardization and traceability provided by blockchain [22].

DLT applications have been proposed for retail trading procedures. Third-party systems that include brokers, exchanges, market-reporting firms, banks, logistics companies, and regulators play a key role in energy markets [23].

From such a grid management point of view, the primary technological obstacle with P2P electricity trading systems is that every device must handle grid conditions and basic economics, along with price and local conditions. (See Figure 8.5.) In turn,

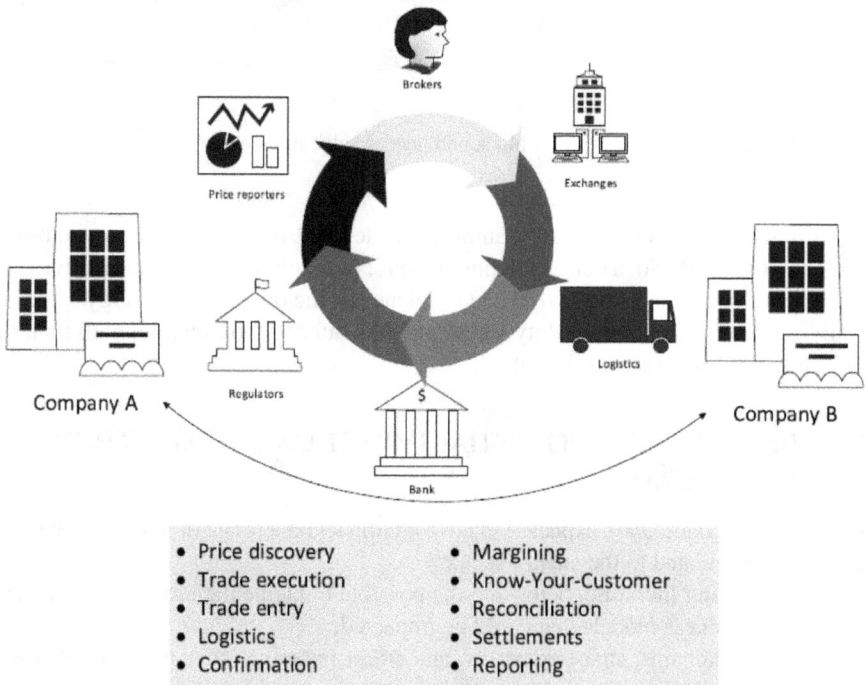

Brokers

Price reporters

Exchanges

Company A

Regulators

Logistics

Company B

Bank

- Price discovery
- Trade execution
- Trade entry
- Logistics
- Confirmation
- Margining
- Know-Your-Customer
- Reconciliation
- Settlements
- Reporting

FIGURE 8.5   Energy composition for the Company A and Company B model.

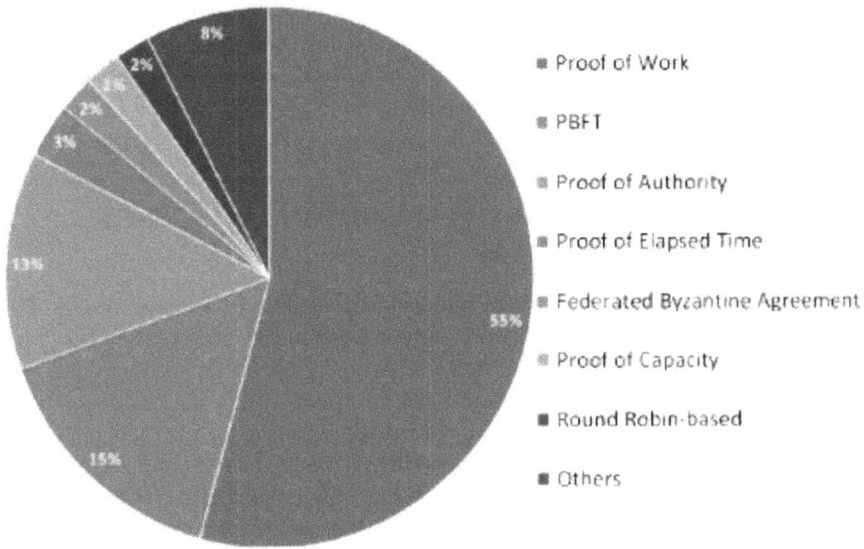

Proof of Work

PBFT

Proof of Authority

Proof of Elapsed Time

Federated Byzantine Agreement

Proof of Capacity

Round Robin-based

Others

FIGURE 8.6   Ratio of operating system techniques comparison.

customer demand would be expected to supply electricity forecasts to the device operator, as already done in existing energy markets. At large scales, machine learning can help predict the potential actions of power users.

Several organizations are working on new blockchains, including water and thermal meters. Blockchain infrastructure is in a networked metering system that monitors water use on a distributed system. They try to make a machine that can pick up on repairs without notice. It is further hoped that similar applications for electricity and gas will be produced. Engie has recently unveiled a new partnership with Air Products to map and verify the renewable energy used in their production processes by using blockchain technology. Blockchains must be carefully considered when used to manage smart meter data. Blockchains or ledgers are by definition available to all (see Figure 8.6). Maintaining the confidentiality of data may necessitate developing novel techniques which make energy consumption difficult to identify. It can be much more difficult to ensure privacy with a decentralized system such as blockchain, such as the EU General Data Protection Regulation (GDPR). One possible alternative may be the use of pseudonymous or permissioned ledgers, where only those with permission to view the data will be given the data. Another critical problem will be the issue of whether or not to keep a distributed, long-length ledger book.

8.6   WHOLESALE ENERGY TRADING

The aim of a passive optical network (PON) is to create blockchain-based solutions in the energy space for region-specific trading of energy. A blockchain-based network was used to buy and sell power between Priogen and GoByte in 2016. The P2P trading platform has so far partnered with over 40 European energy trading firms and utilities in order to grow regional and time-based physical and day-ahead market trading, as well as trading through an upstream and downstream context [24]. Another significant partnership for PON is the Smart Grid for Flexible, Balancing, and Local Energy Transitions joint effort that is headed by the Norddeutsche Energiewende 4.0 (NEW) [25].

8.7 BLOCKCHAIN TRADING FOR UTILITIES AND ENERGY SYSTEM STAKEHOLDERS

Various programs attempt to open their doors to all sectors of the energy system. This is aimed at developing a digital network built on Ethereum for transmitting operators, manufacturers, and customers, and those who supply energy as well as those who consume. To exchange real-time operational information with stakeholders, energy distribution, metering, and billing will be made decentralized using blockchain protocols (see Figure 8.7). That platform uses artificial intelligence techniques to achieve automated responses and to collect business intelligence. Bitt also combines supply and demand to make business deals in one hour. It is suggested that P2P transactions

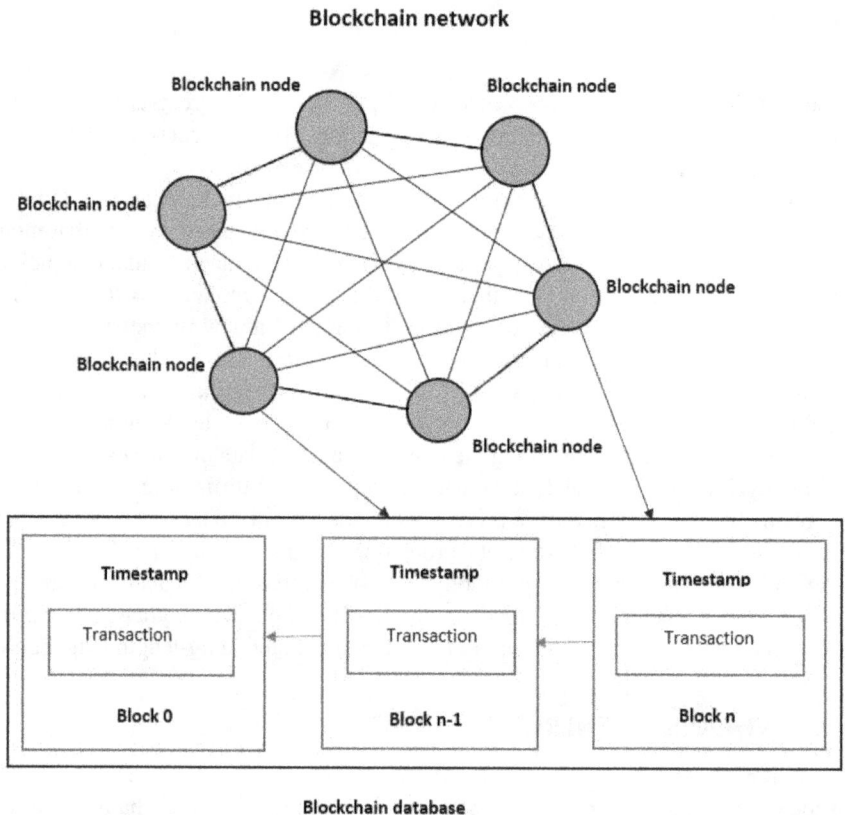

FIGURE 8.7 Database connection on blockchain network.

can be made using Bittwatt's latest cryptocurrency, BWT. It has built a distributed network where stakeholders (generators, utilities, regulators) can exchange real-time price information in a credible manner [26].

8.8 CONCLUSION

Blockchain is a promising technology for a trusted measurement and monitoring of energy-related assets, which are expected to be more decentralized as self-generation and microgrid business scenarios spread, in both the public and private sectors. Much research remains to be done both on the theoretical foundations of blockchain and on the relevant business applications for energy management. Smart contracts and peer-to-peer transactions, as the basis and important component of regional energy trading models, will remain hot topics in blockchain and energy research in the future. The application of blockchain technology in the field of renewable energy, electric vehicle charging, and shared charging infrastructure will always be the focus of blockchain and energy research.

REFERENCES

[1] Zheng Z., Xie S., Dai H.-N., Wang H. *Blockchain challenges and opportunities: A survey.* https://www.henrylab.net/wp-content/uploads/2017/10/blockchain.pdf, [accessed 5 Jun 2018] (2017).

[2] Froystad P., Holm J. *Blockchain: Powering the internet of value (White paper).* https://www.evry.com/globalassets/insight/bank2020/bank-2020-blockchainpowering-the-internet-of-value-whitepaper.pdf, [accessed 18 Jun 2017] (2016).

[3] Baliga A. *Understanding blockchain consensus models.* https://pdfs.semanticscholar.org/da8a/37b10bc1521a4d3de925d7ebc44bb606d740.pdf, [accessed 29 Jun 2017] (2017).

[4] Hyperledger. *Hyperledger architecture, volume 1: Introduction to hyperledger business blockchain design philosophy and consensus.* https://www.hyperledger.org/wp-content/uploads/2017/08/Hyperledger_Arch_WG_Paper_1_Consensus.pdf, [accessed 20 Nov 2017] (2016).

[5] Ethereum Wiki. *Proof of stake FAQ.* https://github.com/ethereum/wiki/wiki/Proof-of-Stake-FAQ, [accessed 24 Oct 2017] (2017).

[6] Ethereum Wiki. *On sharding blockchains.* https://github.com/ethereum/wiki/wiki/Sharding-FAQ, [accessed 8 Jun 2018] (2018).

[7] Bentke J. *On-chain vs off-chain.* https://energyweb.atlassian.net/wiki/spaces/EWF/pages/17760291/On-Chain+vs+Off-Chain, [accessed 8 Jun 2018] (2018).

[8] Reese A.A. *Survey of second layer solutions for blockchain scaling, Part 1.* https://www.ethnews.com/a-survey-of-second-layer-solutions-for-blockchain-scalingpart-1, [accessed 8 Jun 2018] (2018).

[9] Poon J., Dryja T. *The Bitcoin lightning network: Scalable off-chain instant payments.* https://lightning.network/lightning-network-paper.pdf⟩, [accessed 29 Aug 2018] (2016).

[10] Swanson T. *Consensus-as-a-service: A brief report on the emergence of permissioned, distributed ledger systems.* http://www.ofnumbers.com/wp-content/uploads/2015/04/Permissioned-distributed-ledgers.pdf, [accessed 21 Jul 2017] (2015).

[11] Pilkington M. *Blockchain technology: Principles and applications.* https://ssrn.com/abstract=2662660, [accessed 18 Jun 2017] (2015).

[12] Mazieres D. *The stellar consensus protocol: A federated model for Internet-level consensus*. https://www.stellar.org/papers/stellar-consensus-protocol.pdf, [accessed 1 Feb 2018] (2015).

[13] Energy Web Foundation. *Energy web foundation*. http://energyweb.org/, [accessed 24 Nov 2017] (2017).

[14] Intel Corporation. *Sawtooth introduction*. https://sawtooth.hyperledger.org/docs/core/releases/latest/introduction.html#, [accessed 4 Nov 2017] (2017).

[15] Buntinx J.P. *What is proof of elapsed time?* https://themerkle.com/what-isproof-of-elapsed-time/, [accessed 4 Nov 2017] (2017).

[16] Hertig A. *Intel is winning over blockchain critics by reimagining Bitcoin's DNA*. https://www.coindesk.com/intel-winning-blockchain-critics-reimaginingbitcoins-dna/, [accessed 4 Nov 2017] (2017).

[17] Li X., Ma E., Qu H. Knowledge mapping of hospitality research - A visual analysis using citespace, *Int. J. Hosp. Manag.* 60 (2017) 77–93.

[18] Chen C. Citespace II: Detecting and visualizing emerging trends and transient patterns in scientific literature, *Wiley Online Library* 57 (2006) 359–377.

[19] Peng R.-Z., Zhu C., Wu W.-P. Visualizing the knowledge domain of intercultural competence research: A bibliometric analysis, *Int. J. Intercult. Relat.* 74 (2020) 58–68.

[20] Fang Y., Xu H., Jiang J. A survey of time series data visualization research, *MS & E* 782 (2020) 022013.

[21] Mylrea M., Gourisetti S.N.G. *Blockchain for smart grid resilience: Exchanging distributed energy at speed, scale and security*. In: *Proceedings of the Resilience Week (RWS) 2017, IEEE*, 2017, pp. 18–23.

[22] Mengelkamp E., Gärttner J., Rock K., Kessler S., Orsini L., Weinhardt C. Designing microgrid energy markets a case study: The Brooklyn Microgrid. *Appl. Energy* 210 (2018) 870–880.

[23] Mylrea M., Gourisetti S.N.G. Cybersecurity and optimization in smart 'autonomous' buildings. In: Lawless W., Mittu R., Sofge D., Russell S., editors. *Autonomy and Artificial Intelligence: A Threat or Savior?* Springer International Publishing; (2017). pp. 263–294.

[24] Mustafa M.A., Cleemput S., Abidin A. *A local electricity trading market: Security analysis*. In: *Proceedings of the IEEE PES Innov Smart Grid Technol Conference Europe, IEEE*, 2016, pp. 1–6.

[25] Jansen J., Drabik E., Egenhofer C. The disclosure of guarantees of origin: Interactions with the 2030 climate and energy framework, CEPS Special Report, N° 149, 2016.

[26] Awasthi D. Barter to bitcoin: The changing visage of transactions, *Elk Asia Pacific Journal of Finance and Risk Management* 6 4 (2015) 64–71.

9 Blockchain and Government

Anupama Sharma, Ruchi Gupta, Jitendra Kumar Seth, and Seema Garg

CONTENTS

DOI: 10.1201/9781003107507-9

9.1 INTRODUCTION

Digitization and making a ledger of government data through blockchain is a transformative force in various government operations [1]. Blockchain has the potential to assist various government activities and adoption as the technology is continuously evolving. Blockchain-based government [2] documents help to preserve data, make operations fair, and reject double dealing, while increasing its certitude and accountability. Blockchain-based distributed ledgers for government data represent an efficient way to share resources securely using cryptography. A blockchain-based model constitutionally protects the data of a country's citizens held with the government.

Governments get the following advantages through blockchain-based ledger of government data:

- Unshakable retention of government and wary citizen information
- Government efficiently working without intermediary
- Better law enforcement
- Building of trust between government and citizens
- Lowering extravagant expenditure associated with maintaining accountability
- Decreasing corruption
- Increasing transparency in government operations
- Building trust in online civil systems

Figure 9.1 depicts some of the direct benefits of blockchain in many areas, whether government or other areas. Government also may take benefits of blockchain-assisted emerging technologies, such as IoT and artificial intelligence (AI). A number of government and public sector applications, including digital currency/payments, land registration, identity management, supply chain traceability, healthcare, corporate registration, taxes, and voting can be leveraged to support the distributed ledger format. Thus, government activities can be organized better using blockchain and other emerging technologies to serve the nation. Figure 9.2 represents some of the major roles of blockchain usages for government processes.

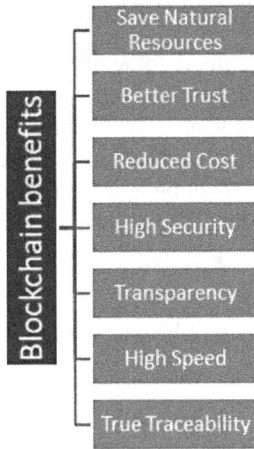

FIGURE 9.1　Benefits of blockchain.

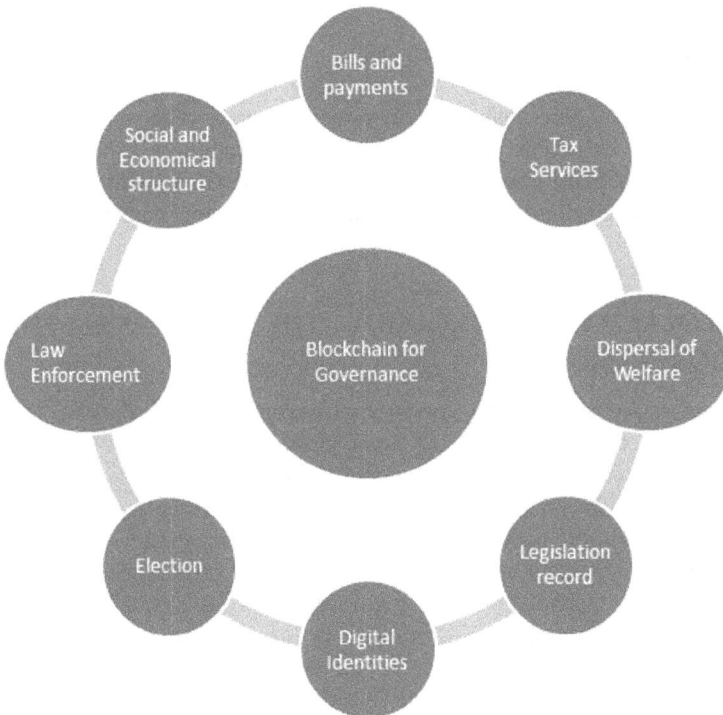

FIGURE 9.2　Blockchain in governance.

9.2   BLOCKCHAIN'S UNIQUE FEATURES

Blockchain technology is made up of some of the technological components which include hash function, cryptographic algorithm, and decentralized network. **Distributed or decentralized** ledger offers evenly distributed power compared to central authorities. These decentralized ledgers are **immutable and auditable**, since immutable means stored data cannot be erased or altered. The hash function is used to avoid alternation in stored data. The cryptographical algorithm which is used in blockchain is a public key cryptography.

9.3   PURPOSE OF BLOCKCHAIN-BASED GOVERNANCE

Adopting block chain technology for government operations will help to begin a cycle of confidence in the judicial and fiscal system that has been lacking in recent years. A few of the domains being affected by blockchain technology are listed below:

- Property Registration
- Identity Administration
- Voting System and Election Process
- Protection of Sensitive Data
- Tracing Supply Chain Management.

The main purpose of blockchain-based governance is to serve the nation better. It may be possible only with building trust of civilians in government activities. True traceability, transparency, identity management, unaltered data on distributed ledger, welfare distribution, handling of tax issues, etc. play an important role managed by blockchain.

9.4   BLOCKCHAIN TECHNOLOGY – THE WAY TO ASSIST THE GOVERNMENT

Blockchain assists government operations to properly manage its key functions. Although blockchain, a decentralized public leader, differs from the governance fundamentals, which are mostly based on centralized public sector units, the incorporation of this technology has the potential to revolutionize the way the government conducts its affairs and make it quick, efficient, and transparent.

The Indian government has announced a spate of projects in recent years that use this revolutionary technology for its e-governance projects. Interest in blockchain has even been shown by state governments. In June 2017, the Andhra Pradesh and Telangana state governments indicated that they were planning to adopt blockchain solutions for land registry, data protection, and KYC (Know Your Customer) records.

Interest in the adoption of blockchain solutions has also been expressed by states like Maharashtra, Gujarat, and Uttar Pradesh. The three key areas listed below are where the inclusion of blockchain in governance can make a fundamental difference to the way the government implements its social agenda.

9.4.1 Making the Provision of Welfare More Efficient

For governments around the world, taking care of the vulnerable and the socially deprived is one of their essential obligations. Welfare schemes, such as old age pensions, scholarships for the disabled, subsidized LPG (liquefied petroleum gas), as well as electricity are the only means of livelihood for the poor and the vulnerable in a country like India, where millions are below the poverty line. Under the Direct Benefit Transfer program, most of these payments can be covered. However, there remains some difficulties in the process, such as fraud, human error, the presence of some applicants who do not have bank accounts, and difficulty prioritizing those who are most in need of government assistance.

9.4.2 Transaction Costs

These can be reduced significantly with the implementation of blockchain technologies by the government. The chances of fraud, leakages, and human error will be considerably lower as the need for third parties to make transactions and maintain records will be minimized.

9.4.3 Land Registration

Vague land titles in India today are a major cause of concern, leading to prolonged litigation. It is possible to make them permanent and tamper-proof by integrating land deeds/records into the blockchain.

9.4.4 Food Value Chain

The government will ensure traceability and accountability in the agricultural supply chain through blockchain technology. It is possible to deal successfully with problems related to unequal pricing, overdue payments, and middlemen.

9.4.5 Ensuring Basic Government Foundation

In current occasions when the danger of an assault is prominent, it is very feasible for individuals with an odious plan to seize control of basic switches. Accordingly, basic information of government associations and organizations is at risk of being seized. Besides, as innovations are additionally used by rail lines, flood boundaries, energy establishments, and many others, the danger of accidents and assaults to harm property and humans increases. With a dispersed record, it is conceivable to screen the trustworthiness of the product for any unlawful changes. It additionally guarantees that information communicated from frameworks with an IoT isn't altered.

9.4.6 Crop Insurance

Blockchain is an ideal stage for regulating crop protection. In its present structure the Pradhan Mantri Fasal Bima Yojana (PMFBY) is managed in bunches. Utilizing the blockchain arrangement, the public authority can add value addition to its services – for example, transferring warnings and rules concerning harvests and limit yields in each instance. Similarly, the insurance agency can also transfer the rates for all harvests in that bunch, utilizing the expert agreement for the group.

9.4.7 BLOCKCHAIN ECONOMY [3]

Blockchain is capable of implementing transactions in an autonomous model. Decentralized autonomous organizations are governed by smart contracts through blockchain, thus making the whole system quite efficient and economical.

9.5 EXPEDIENCY OF BLOCKCHAIN IN GOVERNANCE

Numerous legislatures around the globe are not designed for accepting digital currency; but they do comprehend the significance of blockchain. The public authorities can use blockchain for multiple tasks to enhance day-to-day performance. The advantages of blockchain in government incorporate the accompanying aspects.

9.5.1 DIGITAL IDENTITY MANAGEMENT [4]

The management of digital identity management is a key plan of blockchain technology to process organizations' personal data and to connect various customers' databases. All these processes are not required to store data by involved individual organizations. Moreover, databases are not to be centralized to connect and be accessed from everywhere.

9.5.2 ELECTIONS ON DISTRIBUTED LEDGERS

Blockchain technology provides secure, distributed, and auditable record keeping. This concept may be used during the process of voting in an election to avoid fraud. Moreover, the counting of votes becomes automatic and provides a straightforward decision without distortion.

9.5.3 BLOCKCHAIN FOR CAPITALIZATION

Financial transactions can be protected through blockchain technology to make them secure. Blockchain-based public ledger may be accessible by members having no rights to modify, thus protecting the transaction data and preventing fraud. The whole of the database will be secured once stored with proper encryption. Financial plans of the government can be maintained without loss of data and this reduces the costs of paperwork. It enhances productivity and provides straightforward outcomes.

9.5.4 TRUST BUILDING [5]

Blockchain-governed applications are currently growing in many sectors like banking and healthcare, due to the resulting improved speed and efficient implementation.

The Pew Research Center indicated that the trust of citizens in the American government is almost at a record-breaking low of 18%. Public trust can be enhanced through the use of blockchain technology.

The arrangements governed by blockchain are straightforward and decentralized. For instance, the administrations of Sweden, Estonia, and Georgia are implementing aspects with blockchain-based land vaults, empowering numerous gatherings to safely hold duplicates of the library. This model could help in rapid resolution of debates over property or forestall them in a transparent manner. At the point when citizens and the government share admittance to records, potential for doubt diminishes. These activities improve the trust of citizens in the government processes.

9.5.5 STRAIGHTFORWARD GOVERNANCE

Government organizations can achieve smoother functioning through using blockchain technology, involving lower repetition, dwindling review problems, and enhanced security. Government authorities may have better coordination over this technology. The GSA FAST_Lane measure framework is used to oversee recommendations from sellers; it currently takes 40 days to deal with an approaching proposition, while by adopting the blockchain the time can be reduced to 10 days. Also, GSA officials feel that expenses have been reduced by approximately 80% by adopting the technology.

9.5.6 TRUE TRACING

Blockchain provides true tracing of merchandise which is not possible with conventional tracing systems. The network of production empowered with blockchain turns out to be very efficient. It deals with various networking issues efficiently and makes further follow-ups quite convenient and realistic.

9.6 CURRENT STATUS OF BLOCKCHAIN USAGE BY GOVERNMENTS

Business landscapes have experienced lots of amendment through blockchain in various countries. The steps toward digitization of the international economy will support the development of economic growth. Several countries are actively working on this concept as briefly described below.

9.6.1 CHINA

M/s Alibaba Incorporated is a huge e-commerce company from China which has built an international reputation and plays an important role in China's economy. If we talk about China's economy, Alibaba uses blockchain for cross-border traceability in its operations. It also provides blockchain as a service (BaaS), where users learn to work with blockchain. BaaS provides a platform for organizations where they focus on innovation, not on infrastructure. BaaS by Alibaba offers active support for Hyperledger Fabric, Ant Blockchain, and Quorum. Moreover, Alibaba uses blockchain technology for charities also, which is also known as 'Charities on the Chain' and comes up with transparency to donors as well as beneficiaries. Immutable records of donations are stored in digital ledgers. Another conglomerate, M/s Tencent Incorporated, is evolving as a logistics platform driven by blockchain by partnering with the China Federation of Logistics and Purchasing. China is becoming a blockchain innovation hub as one of the world's biggest economies.

9.6.2 DUBAI

With the help of emerging technologies like blockchain and IoT, Dubai is becoming a smart city by working on three areas for strategic efficiency; these are government, industry, and globalized leadership.

They made estimations that blockchain-based operations can save approximately 5.5 billion Dirhams in documentation resources only. Further, those saved resources increase the growth of the economy.

9.6.3 USA

US FDA signed an agreement with the IBM Watson Health branch to use the distributed ledger technology of blockchain to keep track of patients' data. This partnership has been built to address the transparency as well as security of the data flow process in health organizations.

The previous US government also allowed the Department of Homeland Security in 2018 to adopt an IoT and blockchain technology to protect sensitive border data.

9.6.4 UK

The UK's FSA has also taken steps toward the adoption of blockchains. Their blockchain-based developments are in the areas of food traceability application through integrated blockchain and land registry market, using this technology to make an efficient buying and selling process and reduced land registration frauds.

The Department of Pensions is also assessing this technology to provide comfort and satisfaction to claimants through the transparent management of their money.

9.6.5 INDIA

The number of blockchain-based projects is growing rapidly in India. Most of the state governments such as Kerala, Andhra Pradesh, and Maharashtra are starting to promote schemes for startups and projects to use blockchain technology. A recent report of the World Intellectual Property Organization envisages India gaining sixth from top place in patents related to blockchain technologies. Blockchain-based applications are ubiquitous; hence fresh explorations are appearing in almost every field, particularly in all the processes of the government.

9.7 BLOCKCHAIN APPLICATION IN GOVERNMENT PROCESSES

Blockchain may be used in various aspects of a political process which helps it acquire a more transparent outlook in the eyes of the citizens. It will also improve the efficient workflow of political decisions. A fair politically based government may have an impact on culture, a country's economy, ethics, social values, happiness index, growth index, and many other aspects. Blockchain creates a platform for policy and decision makers to showcase directly their ideology to citizens and it can foster a strong bond among them, thus maintaining greater confidence of citizens for other stakeholders including politicians (democracies). There follow various activities where government may make use of blockchain technology:

9.7.1 BLOCKCHAIN IN THE ELECTION PROCESS

Blockchain technology provides a platform where citizens may get the opportunity to vote on the distributed ledger. This technology removes the changes of forgery during the voting process and subsequent steps, thus resulting in a transparent selection of their leaders. Moreover, voters' identity will be better protected using this technology by hiding their actual identity. A blockchain-oriented election process can be an efficient and effective process which may enhance the future perspective of good governance.

9.7.2 TRANSPARENT PUBLIC PROCUREMENT

Corruption is one of the main problems of public procurement [6]. The selection of vendors is one of the steps of public procurement which is not transparent to an acceptable level. This process highlights many issues such as financial waste, decreased healthy competition, etc. Moreover, it leaves an impression of being complex and opaque, thus generating disagreements among various groups. With the support of blockchain, governance has a chance to set up a system which resolves the associated risk factors by establishing some tamper-proof transactions and maintaining records through smart contracts [7]. Accountability and transparency will not be hindered with suggested smart contracts.

9.7.3 CORPORATE OWNERSHIP REGISTRY

In many countries, the government has started using blockchain technology to create and use central registries which eliminate forgeries and manipulations. It is expected that possibilities of corruption in various stakeholders, including companies, will be eliminated after using this emerging technology. It will directly impact the vested interests such as money laundering, receiving and sending bribes, and using government resources for unethical usage.

9.7.4 GRANT DISBURSEMENTS

The efficiency of a political party is usually judged through its working culture in various fields, such as the education sector, social work, construction, different levels of assistance, etc. Blockchain-based implementation of services may increase trust related to the whole of the involved process [8]. This technology has the capability of reducing the intermediaries' involvement in the activities to grant permission and release funds to the real beneficiaries for the relevant activities.

9.8 GOVERNANCE THROUGH BLOCKCHAIN – MAJOR ISSUES AND CHALLENGES

Blockchain technology is widely accepted in every field of work and governments are preparing to take advantage of it [9]. A technology may possess another side to the coin; also the blockchain may have some uncontested issues and challenges. These issues though nascent may hinder the prevailing benefits of blockchain for government activities. Some of such issues and challenges are described briefly as follows:

9.8.1 INITIAL IMPLEMENTATION COST

Initial cost for implementation of blockchain for a particular system requires proper set up for the technology. There may be initial high cost of development of the technology as well as the maintenance cost which have to be borne by the user departments.

9.8.2 NASCENT STATE OF TECHNOLOGY

As blockchain is an emerging technology, there is a lack of awareness about the technology. Even the implementing departments may face human resource constraints while working on this technology. At present the numbers of blockchain developers are meager in the world. The requisite experts of blockchain technology are in short supply, at least in India.

9.8.3 RISK TO PRIVACY AND DATA BREACH

Though blockchain data is secured through the hash function and public key cryptography, there still remain chances of cyber-attack on the public keys.

There are many other issues and challenges in adopting blockchain technology, such as social repercussions of decentralization of the processes, immaturity or poor decision making during blockchain-based governance [10]. However, a closer look highlights the obvious drawbacks of any technological usages, even of blockchain technology. Also, these problems are not specific to the implementation of blockchain for the government. Further, the blockchain technologists are striving to find solutions to associated problems.

9.9 BLOCKCHAIN USE CASE STUDY IN INDIA

In India, the apex policymaker body of central government is called NITI Aayog [11] which has analyzed the impact of implementing blockchain on various government processes and developed a case study. It has highlighted the presence of many obstacles in the implementation of such systems. Some of the issues are discussed below.

9.9.1 LAND RECORDS

The conventional method of acquiring and maintaining land records and transfer of land ownership is not found to be efficient because many of the departments may be involved in the said process. This involves issues such as land data not being properly recorded in files, poor coordination among various departments resulting in delay in implementation, natural calamities cause lossing of documents, or disputes of ownership on land records. To keep track of land records is a difficult task, making it an inefficient process. Blockchain has been found to be a new way to keep the records on distributed ledgers and transfer of ownership will be easy with this technology. Distributed ledger is immutable and readily auditable, hence, may be efficiently used to record land registries.

The NITI Aayog has found many complexities in adopting this technology for land registry which is a complex process. Major concerns needing to be resolved before transferring the records on distributed ledger are discussed below:

9.9.1.1 High Cost of Litigation

A very large number of disputes have been registered in courts regarding land issues, such as ownership conflicts, title of the properties, cost of the properties. There are many judicial and administrative forums where such unresolved cases are pending. Those needing resolution before getting transferred to blockchain may take more than 10 years to resolve.

9.9.1.2 Establishing Land Ownership

There are many ways to get land ownership; it may be gifted, transferred based on inheritance, purchased, etc. To verify the details of ownership several documents (property tax slips, electricity bills, etc.) are required to be verified. Moreover, all the judicial cases of land ownership need to be completed before transferring to the new system.

9.9.1.3 Asynchronicity of Information

Various agencies are involved in land registries such as Estate and sub registrar offices; hence, many cumbersome pre-processes need to be completed to make a synchronous system.

The NITI Aayog has completed a survey on the use of blockchain-based governance in the Union Territory of Chandigarh, and highlighted the need for the 'digitization' of processes to enhance the usage of existing IT systems. Blockchain features are analyzed, being critical to execute 'smart contracts' and to simplify the involved government processes.

In general, blockchain-enabled systems are able to create immutable records of land ownership. These records are digitally stored for the future with the ability to audit from anywhere and at any time. This technology is highly recommended and is being adopted in governance in many countries.

9.9.2 SUPPLY CHAIN MANAGEMENT FOR PHARMACEUTICAL DRUGS

One of the worldwide issues is related to spurious drugs. The Ministry of Health and Family Welfare in India performed a national drug survey in 2016. WHO has made an estimation on the basis of the national drug survey that 1 out of 10 medicines are fake and this needs immediate attention. Fake drug production is a malpractice which leads to significant risk for patient health.

Numerous benefits have been observed in using blockchain for pharmaceutical supply chain management, as it enhances transparency and reliability of transactions in the pharmaceutical industry. Blockchain allows manufacturers to access real-time data. Moreover, consumers will be able to verify the drug provenance on its purchase.

Thus, major benefits of blockchain are as follows:

- End-to-end optimized tracing of pharmaceutical drugs: Drugs in transition may be tracked at any point of shipment through blockchain. True tracing is required to fulfill business commitments.
- Enhanced accountability through transparency: This facilitates the efficient management of inventory and allows batch reminders which are required for patient health safety.

Blockchain technology will be able to remove the dependency of intermediate agents in the drug supply chain. Efficient supply chain management with quality control will enhance the reputation of this industry.

The case study [11] of using blockchain by the NITI Aayog includes many other issues and suggested some solutions such as digital identity management through blockchain to remove fraud in educational certifications, blockchain for legal platforms, blockchain for insurance, blockchain for organic farming, blockchain for energy trading.

9.10 BLOCKCHAIN AND GOVERNMENT – FUTURE GOVERNANCE

Blockchain technology has much potential to establish a fruitful political system which may remove the loopholes of the current system [12]. The efficient use of national resources will be implemented and tracked transparently through blockchain-based distributed ledger technology. Smart contract will streamline the vendor selection process without any fraud and involvement of vested interests.

Blockchain-based governance is expected to be totally transparent and accessible to civilians too. The major change will be toward a transparent process, since distribution and utilization of government funds can be analyzed by the stakeholders in the system.

9.11 CONCLUSION

Blockchain and governance is a big milestone to achieve, particularly in developing countries. Governments of developed countries, namely the USA, have already taken steps toward implementing emerging technology to take its flavor in various activities and processes of governance.

Major outcomes of blockchain technology-assisted governance are transparent processes to gain better trust of citizens, fast and efficient execution of government processes, law enforcement, and procurement, to name a few. Government assets can be traced properly through blockchain-based distributed ledgers. The systematic organization of various government departments is possible under blockchain-based governance. The government is responsible for maintaining a social, ethical, secure, and progressive environment for civilians. Blockchain-based governance will be better able to maintain such a secure environment and evenly prioritized processes for the betterment of country people. Blockchain-based governance is the future governance to bring about a revolution in politics. Polities without any fraud and unethical steps will be possible through this technological change. All third-party involvement for their interests in various political matters will be removed. Political leaders who want to serve the nation will get a platform to work fear-free. Their decisions and fair execution will be visible to everybody on public ledger. Smart contracts play an important role in removing forgery from various procurement processes. Blockchain-based government is the only way to serve the nation in an efficient manner where natural resources are saved while enhancing different activities. It will become a game changer for future generations to bring efficient governance and that is the requirement of current self-disciplined societies.

REFERENCES

[1] Arruñada, B., & Garicano, L. (2018). Blockchain: The birth of decentralized governance. *Pompeu Fabra University, Economics and Business Working Paper Series, 1608*.

[2] De Filippi, P., & McMullen, G. (2018). *Governance of blockchain systems: Governance of and by Distributed Infrastructure* (Doctoral dissertation, Blockchain Research Institute and COALA).

[3] Beck, R., Müller-Bloch, C., & King, J. L. (2018). Governance in the blockchain economy: A framework and research agenda. *Journal of the Association for Information Systems, 19*(10), 1.

[4] Lesavre, L., Varin, P., Mell, P., Davidson, M., & Shook, J. (2019). A taxonomic approach to understanding emerging blockchain identity management systems. *arXiv preprint arXiv:1908.00929*.

[5] De Filippi, P., Mannan, M., & Reijers, W. (2020). Blockchain as a confidence machine: The problem of trust & challenges of governance. *Technology in Society, 62*, 101284.

[6] AlShamsi, M., Salloum, S. A., Alshurideh, M., & Abdallah, S. (2021). Artificial intelligence and blockchain for transparency in governance. In *Artificial Intelligence for Sustainable Development: Theory, Practice and Future Applications* (pp. 219–230). Springer, Cham.

[7] Murray, A., Kuban, S., Josefy, M., & Anderson, J. (2019). Contracting in the smart era: The implications of blockchain and decentralized autonomous organizations for contracting and corporate governance. *Academy of Management Perspectives*, (ja).

[8] Paech, P. (2017). The governance of blockchain financial networks. *The Modern Law Review, 80*(6), 1073–1110.

[9] Rikken, O., Janssen, M., & Kwee, Z. (2019). Governance challenges of blockchain and decentralized autonomous organizations. *Information Polity, 24*(4), 397–417.

[10] Ziolkowski, R., Miscione, G., & Schwabe, G. (2020). Decision problems in blockchain governance: Old wine in new bottles or walking in someone else's shoes?. *Journal of Management Information Systems*, *37*(2), 316–348.

[11] https://niti.gov.in/sites/default/files/2020-01/Blockchain_The_India_Strategy_Part_I.pdf

[12] Katina, P. F., Keating, C. B., Sisti, J. A., & Gheorghe, A. V. (2019). Blockchain governance. *International Journal of Critical Infrastructures*, *15*(2), 121–135.

10 Blockchain in Healthcare
A Smart and Hierarchical Decentralized Tracing Application in Healthcare by Blockchain

M. Kumar, Himanshu Dubey, and K. Kumar

CONTENTS

10.1 INTRODUCTION

Blockchain technology supports a distributed and hybrid milieu with no requirement for a principal authority. While scripting, a keen focus toward the blockchain has centered on cryptocurrency, mainly "Bitcoin2" and the impact of blockchain is expected to have on the financial sector. This result is directed to the well-known consultancy, "labeling" it as one of three skills that will transform the fiscal services world.

Notwithstanding the attention on fiscal facilities, there are numerous other zones for movement in real estate, supply chain management, and healthcare. Healthcare is the major for commotion due to the diversity of glitches in the business which can be resolved by blockchain through its immutability, fraud prevention, and capability of sharing data between hospitals without requiring trust.

The existing complexity within the current healthcare system is recorded in Table 10.1. A main complexity, as known by Frost & Sullivan, to tag the health apparatus along with the operable ID and in amalgamating belief in maneuvering recognition and tracing. When a maneuver, such as a distillation pump, is revealed to consume crashed, the chasing of the maneuver can disclose the basis of the complexity and stop needless repurchasing in the case of a missing maneuver. A robust belief set up created nearby the recognition of remedial maneuver is possible to decrease

DOI: 10.1201/9781003107507-10

TABLE 10.1

Existing Complexity within the Current Healthcare Industry

Domain Issue	Description
Healthcare Data Interchange	Information must be clear between healthcare workers and required third parties, brokers, and patients, while constantly observing data protection in the healthcare system.
Nationwide Coordination	Consuming a sole standard for patient information interchange allows ease of transient information between healthcare workers, which request systems regularly do not offer.
Medical device tracking	Healthcare maneuver tracing from sources chain to discharging permits for rapid recovery of devices, anticipation in needless purchasing, etc.
Drug tracking	In the context of healthcare, blockchain provides the competence to trace the chain of care from supply chain to patients and also permits the rapid detection, and restriction, of fake medicines.

these coercions. The information approximates that only 20% to 30% of health maneuvers are associated within infirmaries, due to safety,and secrecy complexity. In the curative industry, blockchain can relieve the cumulative jeopardies about fake and non-approved drugs. With the maneuver tracing, it is workable to explain "smart contracts" for painkillers and then classify pill ampules, with unified "GPS" and chain-of-custody classification.

In clinical tests, blockchain can be utilized to stun the complexity of fake outcomes and remove information, which does not keep the bias of authors or backing source's purpose. This will apply truthfulness in clinical tests.

10.2   LITERATURE REVIEW

In the paper [1] the authors discuss the Internet of Things (IoT) and also the advances in IoT which leads to the new term Internet of Computing (IoC). It also informs about the combined use of the IoT and IoC in the many sensor technologies in the upcoming times.

In the paper [2] the authors discussed the use of cryptography and steganography techniques for security purpose in the IoT. In this paper the elliptic variety galois cryptography protocol is used in which a cryptographic technique is used to encode the secret information and then uses the matrix XOR coding steganography technique to embed the secret message; then the adaptation firefly algorithm is used to optimize the canopy blocks among the images, and the information which is hidden is recovered and decrypted.

In the paper [3] the authors discussed about the use of cyber security in the distributed and parallel computing system and use of the blockchain in the cyber security system which can enhance the security of the information system.

In the paper [4] the authors suggest a blockchain framework in medical which includes all shareholders in the healthcare system to examine prospects and contests by giving a unified blockchain design.

In the paper [5] the authors elaborate the combination of data, network, and blockchain technology in healthcare. Healthcare data have become an imperative element in this sector. There is a need for the secure and safe transmission of medical data among the organization. The big challenge is to collect the information about patient diseases which has become essential to restrict the pandemic in the world. There are numerous ways of doing this, although it is not possible to obtain complete and temporal information at a given time. To bring transparency between hospitals around the world, there is a need to share the data directly across the world. This can only be done via technology which includes a combination of IoT networks and blockchain which is decentralized in system and has the capability to store the data in a distributed manner.

In the paper [6] the authors review the use of blockchain technology in the healthcare system. The use of this technique is increasing day by day and there are various areas within the healthcare system where blockchain technique could make a huge impact which could enhance our healthcare system.

In the paper [7] the authors proposed the programmable blockchain method to overcome the challenges in the healthcare system such as the communication gap, inefficient clinical report delivery, and poor management of health records. It can also provide the evaluation metrics to assess the blockchain-based distributed applications with respect to their feasibility, capabilities, and healthcare system compliance.

In the paper [8] the authors proposed to upsurge the consideration of blockchain technology as a data store and to endorse a systematic method for bulky software systems. In this paper, authors classify the mutual layers of a typical software system with data stores and conceptualize each layer in blockchain terms. The authors also inspect the location and movement of information in "blockchain-based" operations. Finally, the authors scrutinize the information administration complexity in blockchain with relationship assurance in terms of privacy and security.

In the paper [9] the authors discuss the potential usage of blockchain technology in administration, with a summary of the latest advances in the arena of blockchain technology as an instance of methodical maturity and administration. The association between Individual(s), Right(s) and Entity(s) in a hospital system is the foundation for the description of mandatory functionality, assuming the intricacy between these three fundamentals: uniqueness of an individual, lawful variety in objects. The paper investigates certain principles of good administration, which include transparency, accountability, security, and regulation, made possible by blockchain technology. It is concluded that the method is not adequate for operation in administration at the moment.

In the paper [10] the authors describes the blockchain and its dispersion looks to be changeable for dissimilar industries. The objective of this study is to discover the blockchain technology dispersal in various industries through a mixture of academic literature and social media.

10.3 PROPOSED FRAMEWORK

The health-tracking structure has become vital for delivering health reports within a definite time. The health-tracking system can track the health of human beings with three-tier architecture through which people interconnect to the IoT network and hospital management system that contains the artificial intelligent system to attain the information based on the identity of people that is needed to create the chain of a particular patient and detect the health issues based on patient history with a phenomenon of blockchain technology where it is needed to get the data request of the patient and transfer that information to all existing nodes in the system for cloning the information and storing the information in a distributed manner. The distributed information helps the availability of information at any place from where want to access. Blockchain architecture recommends a decentralized system as an environment of distributed node that stores information of a particular person and provides in the chain whenever required. The benefit of the architecture is distributed in nature since the failure of a single node does not lean to a failure of all node information, while the crashing of a single node will not crash the whole system. In this paper, the authors describe a tracking system of a disease and how a pandemic can be restricted by transparency tracing via blockchain by inclusion of another learning concept called machine learning that helps to evaluate the pattern.

The disease influenced patient required to be ascertained as timely as possible and the analysis of a sizable amount of people can be achieved at their inland since it is spread from human to human and necessary to protect individual. There is a prerequisite for a block system that formed a system for taking patient data to test and analyze at their inland and can deliver the outcomes instantly. There are numerous techniques existing that are useless in the particular, which carries us about to improve a domain-specific study for evolving a system. There are numerous health expert systems existing to assess and foresee ailments based on the information delivered by the patient (Figure 10.1).

The framework in the above figure shows how to develop an expert system based on blockchain technology, in which a chain of connected nodes carries the request from the users and distributes it in a distributed environment where each node contains the copy data for validation and verification purposes. This helps in tracking the health of a patient whose history is maintained in a distributed database. In this way, a health-tracking system is formed for tracking the health of patients.

Hence, this system is based on blockchain and is known as a healthcare tracking system for tracking the data of patients who are under the control of a particular doctor, but the patient data would be distributed among several authentic labs over the world.

10.4   DATA FLOW ANALYSIS IN BLOCKCHAIN

In which the data flow in architecture of blockchain is shown via the hospitals' system in an associated network of hospitals worldwide is shown, which consists of events that are linked among them and functioned, which usually shows the flow of the data in the network of a hospital through the blockchain for transparency and tracing about the pandemics.

Figure 10.2 shows the diagram of blockchain in hospitals, in which the flow of the events through the system in hospitals is shown which consists of events that are associated among them and are worked and linked in the hospitals, which generally shows the flow of the data in the hospitals through the various processes. As the data stored in the blockchain is immutable. the path taken by the blockchain for data flow from generation to the end is negligible. For a better flow diagram, the necessary

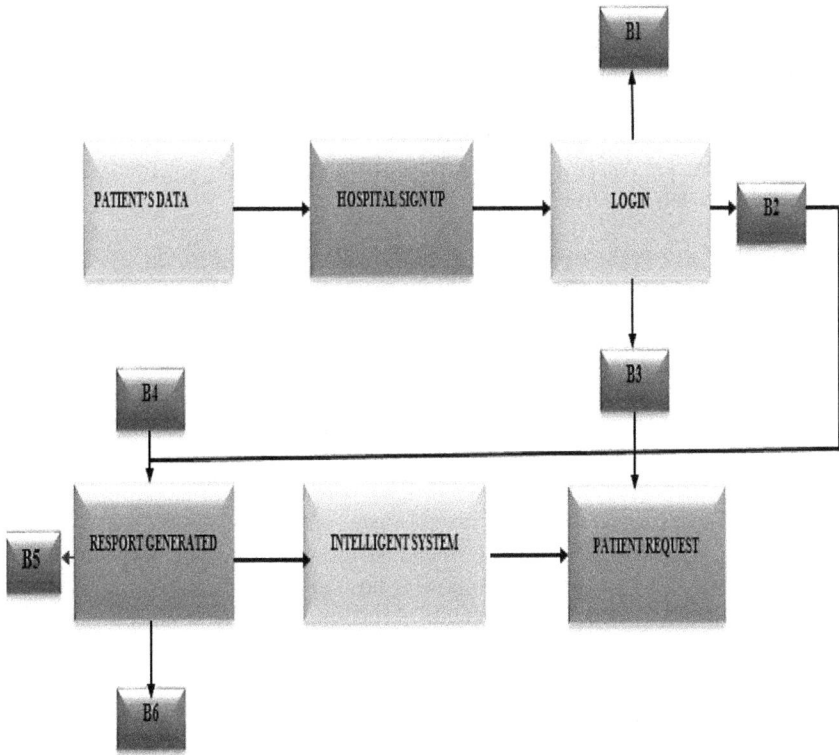

FIGURE 10.1 Proposed framework.

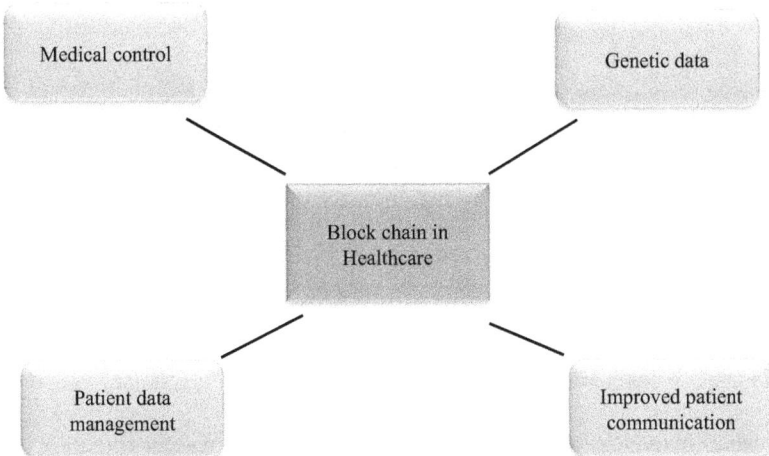

FIGURE 10.2 Blockchains in healthcare.

features of blockchain are present. This refers to the flow of the digital data acquisition in the internal and external databases for the production data transfer. The actual databases after preprocessing and the feedback database are not that much necessary, because the infrastructure in both scenarios is the same, so it can be assumed or neglected. These healthcare technologies are useful and helpful in the blockchain technical production. While looking at the production in the digitization, the chain of an events approach is as follows:

(1) System
(2) Patient data
(3) Processing
(4) Record.

A healthcare system that is based on the blockchain can safeguard data, lessen fraud, and streamline processes, waste, and misuse, while concurrently increasing trust and responsibility. A hospital that is based on the blockchain model, independent businesses and hospitals divide assets over a dispersed ledger, safeguarded by using cryptography. For protecting sensitive citizen and hospital data, the structure given above is used that will also eliminate the single point failures. A hospital that is based on the blockchain has the capability to resolve provision main points and allow the following advantages:

(1) Preserving the patients' and hospitals' data
(2) Diminution of patient-concerted processes
(3) Attenuation of immoderate costs related to managing liability and hospitality
(4) Diminishing abuse and dishonesty about the data
(5) Enhancing belief in the online healthcare system
(6) Building up faith between the healthcare workers and patients
(7) Enhancing efficiency
(8) Safeguarding sensitive and significant data
(9) Maintaining patient indices.

The healthcare workers' distribution ledger style can be grasped to assist a cluster of hospitals, which includes the sign up and login, identification management, healthcare, voting, supply chain management, and land registration. There are numerous hospitals that join the battle to process statutory legislation and begin pilot projects that are centered on blockchain technology. It can leverage blockchain technology to supply process optimization and data security. Hospitals and authority systems have a complex and ever-evolving issue of identity and security. For solving the issues related to the data authenticity and transparency system and identity, blockchain is used because it has exceptional utility to provide the solutions. There are many benefits in using blockchain technologies to improve healthcare systems:

(1) Medical devices will be advanced with a high level of security
(2) The data of patients in hospitals are transparent so they can be easily tracked
(3) Transactions done by hospitals are faster by using blockchain
(4) The chances of hacking threats are also reduced to a larger extent

(5) There is no need to pay centralized services because blockchain has decentralized platforms
(6) Different levels of accessibility are also offered by the organization's blockchain technology
(7) Interoperability
(8) Affirm adjudication
(9) Supply chain management.

10.5 RESEARCH IMPLICATIONS

In the healthcare sector, blockchain technology has emerged as a tremendous potential to develop an approach which is more patient-centric and can connect different systems to trace diseases quickly by getting more precise results in electronic healthcare records. This chapter has discussed an advanced research and analysis in the context of healthcare. The main objective is to get information about existing systems and their limitations and to seek solutions toward blockchain technology in the healthcare sector. The analysis approach consists of collection of data and its properties with a bibliometric review. The conclusion specifies that blockchain technology research in healthcare is cumulative and it is typically consumed for sharing of information, to trace health across the world. Hence, blockchain is playing a vital role in developing a system in the context of healthcare which distributes information about diseases and manages the health of patients by keeping all their records.

- Future work includes the integration of multiple hospital data for early diagnosis of disease.
- To work on a combination of a decentralized system and a resilient distributed data approach with cloud computing to invent a health-tracking system.

10.6 APPLICATION WORK

The research gaps encountered by researchers for positioning of projected and advanced administration system in healthcare are described as:

(1) The upward and huge investigation conceded in this area with parallel methods keep out the authors' new approach.
(2) The lack of accessibility of papers means for investigators that definite studies are absent from the research paper as there are present some articulators that cannot be immediately obtainable.
(3) The rising and enormous request of research in the information security field makes it likely that the authors will leave out some of the principal research from an interconnected study directed by investigators.

10.7 CONCLUSION AND FUTURE SCOPE

In recent years, blockchain technology has become very popular in various contexts, especially due to the attractiveness of cryptocurrencies. In the healthcare sector, blockchain technology has emerged as a tremendous potential to develop an approach which is more patient-centric and can connect different systems to trace diseases quickly by getting more precise results of electronic healthcare records. This chapter has discussed the advanced research and analysis in the field of healthcare. The chapter discussed how existing systems work, and their limitations. The chapter also discussed the solutions in the healthcare sector using blockchain technology.

REFERENCES

1 Jha, S., Kumar, R., Chatterjee, J. M., & Khari, M. (2019). Collaborative handshaking approaches between internet of computing and internet of things towards a smart world: A review from 2009–2017. *Telecommunication Systems*, 70(4), 617–634.

2 Khari, M., Garg, A. K., Gandomi, A. H., Gupta, R., Patan, R., & Balusamy, B. (2019). Securing data in Internet of Things (IoT) using cryptography and steganography techniques. *IEEE Transactions on Systems, Man, and Cybernetics: Systems*, 50(1), 73–80.

3 Le, D., Kumar, R., Mishra, B. K., Khari, M., & Chatterjee, J. M. (2019). *Cyber Security in Parallel and Distributed Computing*. Wiley, Hoboken.

4 Gökalp, E., Gökalp, M. O., Çoban, S., & Eren, P. E. (2018, September). Analysing opportunities and challenges of integrated blockchain technologies in healthcare. In *Eurosymposium on Systems Analysis and Design* (pp. 174–183). Springer, Cham.

5 Farouk, A., Alahmadi, A., Ghose, S., & Mashatan, A. (2020). Blockchain platform for industrial healthcare: Vision and future opportunities. *Computer Communications*, 154, 223–235.

6 Hasselgren, A., Kralevska, K., Gligoroski, D., Pedersen, S. A., & Faxvaag, A. (2020). Blockchain in healthcare and health sciences—A scoping review. *International Journal of Medical Informatics*, 134, 104040.

7 Zhang, P., Walker, M. A., White, J., Schmidt, D. C., & Lenz, G. (2017, October). *Metrics for assessing blockchain-based healthcare decentralized apps*. In *2017 IEEE 19th International Conference on e-Health Networking, Applications and Services (Healthcom)* (pp. 1–4). IEEE.

8 Paik, H. Y., Xu, X., Bandara, H. D., Lee, S. U., & Lo, S. K. (2019). Analysis of data management in blockchain-based systems: From architecture to governance. *Ieee Access*, 7, 186091–186107.

9 Vos, Jacob, Lemmen, Christiaan, & Beentjes, Bert (2017, March). *Blockchain based land administration feasible, illusory or a panacea*. In *Netherlands Cadastre, Land Registry and Mapping Agency. Paper prepared for presentation at the 2017 World Bank Conference on Land and Povertry. The World Bank, Washington, DC.*

10 Grover, P., Kar, A. K., & Janssen, M. (2019). Diffusion of blockchain technology: Insights from academic literature and social media analytics. *Journal of Enterprise Information Management*. ahead-of-print. doi: 10.1108/JEIM-06-2018-0132.

11 Blockchain for IoT Edge Devices and Data Security

Anubhav Singh, Swaroop S. Sonone, Mahipal Singh Sankhla, Kapil Parihar, and Mansi Saxena

CONTENTS

11.1   INTRODUCTION

The quick development in evaluating controllers and devices' correspondence advances is advantageous to important advances in our overall population. This has brought about an augmentation in the quantity of sensible electronic machines for almost zones, a decrease in their creation charges and a change in viewpoint from this contemporary reality into the front-line world. Thus, the methods by which we interface with anybody and with the environment has restored, working late development to obtain a prevalent origination of the globe. As the Internet of Things (IoT) has arisen, a lot of developments as of radio frequency ID (RFID) to wireless sensor networks (WSN), which gives the abilities to spot, instigate with it and talk about it on the internet [1, 7]. Right now, an IoT gadget it might be electric contraption from an abiliment to a device headway stage and the chance of usages where it tends to be misused fuse ample districts of the generally useful public. The IoT has a central influence in changing stream metropolitan regions into insightful metropolitan regions, electrical cross sections into splendid organizations and council into keen homes, as this is already said in the beginning. As per various investigation data, the quantity of related contraptions is foreseen to go someplace in the scope of 20–50 billion till 2020 [2, 8] overwhelmingly in light of the colossal number of devices which IoT can fake it. As the IoT imagines a totally related world, that has things prepared to confer assessed data and help out each other. This additionally makes possible an electronic depiction of this current reality, presumed that the greater part of the canny applications in a combination of undertakings may be made. This involves smart water, wearables, smart homes, smart metropolitan territories, automotive, healthcare, environment, smartgrid, etc. IoT courses of action are being sent in various zones, smoothing out creation and digitalizing organizations. IoT applications have undeniable credits, they produce tremendous volumes of data besides, require organization and power for huge stretches [9, 1].

The limitations in recognition, PC breaking point, associations, and confined power stock address a variety of difficulties. The gigantic augmentation of IoT should be maintained by typical frameworks and programs to decrease the current exhibit in the field. This exhibit prompts directly storage facilities and diminishes the gathering of the IoT. In any case, next to the exhibit and coordination experiences existing in the IoT, the steadfastness of its data is moreover a huge issue to remain in perception [7, 15]. At the present time, we have confidence in the statistics of economic substances and the public authority between others, yet would we have the option to confirm about the data given to them by anyone external substances, for instance, IoT associations, has not been modified/changed/distorted in any way? This is an irksome request to answer in concentrated designs. Untrusted substances can change information consenting for their own potential benefits, so the information they give may not be absolutely strong [8, 16]. This accomplishes the need to affirm that the information has never been changed. One way to deal with give unwavering quality in IoT data is through a scattered assistance trusted by the total of its individuals that guarantees that the data stays perpetual. In case all individuals take the data and by the way which they affirm that the data isn't changed from the foremost portrayal, reliability that is refined [8, 16].

Also, sharing a system that guarantees data constancy would allow organizations to share and securely move information with inhabitants. In various domains which are far reaching obviousness of assets for the duration of their life expectancy is required by rules, data changelessness transforms into a key test [8, 17]. Emphatically, European Union (EU) rules include food producers to follow and perceive each unrefined material used in the elaboration of their food things despite the last unbiased of all of them. For example, by virtue of a gigantic food association with an enormous number of amassing suppliers, besides countless clients, the information ought to be digitized and is taking care of robotized to adjust to rule. One model of strong rule is the pork supply, overseen in various countries. In the present circumstances, despite following the unrefined material used in pig foodstuffs and medicines and the last unbiased of the pork, the transport of the animals in the midst of plants ought to be similarly selected by the guidelines [7, 22].

This kind of circumstance involves various individuals, some of them really relying upon non-automated information dealing with procedures. Because of food contamination, which has been a huge issue for the complete people's prosperity from the earliest starting point of time, information which is lost or difficult to find recommends breaks in the region of the trouble's center interest. Additionally, in like manner achieve the individuals' inquiry concerning soiled things likewise, a colossal decrease to their greatest advantage. As demonstrated by the World Health Organization (WHO), it is evaluated that reliably around 600 million people on the planet experience the evil impacts of eating corrupted food, of which 420,000 fail horrendously from a comparative explanation [3]. In this way, missing or closed off information can impact food supervision and customer prosperity. In this sort of circumstances, the IoT can change and adjust the business and society, digitizing the data with the objective that it might be addressed and handled in a certifiable period. This development can be used to advance existing cycles in various domains, for instance, metropolitan regions, industry, prosperity, and transportation. Notwithstanding the way that the IoT can empower digitization of the information all alone, the reliability of such information is so far a fundamental test [16, 23]. Also, another advance which is imagined as the underlying regionalized computerized cash can offer a response to the data reliability issue: Bitcoin, which has changed the parts in real money moves [24].

Bitcoin computerized cash; additionally, countless its looming varieties, can be throughout the world moved without money-related components and new exchanges, with a high level, non-adaptable wallet. Bitcoin is maintained by a program that nuances the establishment at risk for ensuring that the information stays unchanging after some period. This program is called the blockchain [2, 23]. This applies to various regions, as information perpetual nature is guaranteed that applications should go to the past computerized monetary standards. Blockchain may have vexed reliability of information as well. For instance, this development has been used in fair structures by government components, renting and data amassing among others [4]. This part has the current challenges of blockchain and IoT and the feasible central purpose of their solidified practice is explored. Inconvenient applications are included here, despite a review of the available blockchain stages to address these troubles [5].

11.2 BLOCKCHAIN

The issue of trust in information structures is complicated when no affirmation or audit instruments are given, especially when they need to oversee sensitive information – for example, money-related trades with virtual financial structures. In this particular condition, Satoshi Nakamoto, during 2008 [5] introduced two revolutionary methods of reasoning which has an astounding outcome [9, 16]. The absolute first of these is Bitcoin, a PC-created cryptographic cash which holds its prod without assistance starting any brought together incomparability or money related component. Or then again perhaps, the coin is held taking everything into account and securely by a regionalized payer to payer association of performers which makes an auditable and undeniable association [8, 22]. Besides, whose unmistakable quality has withdrawn altogether farther when contrasted with the cryptographic cash itself, it is a blockchain. Also, blockchain is the framework that licenses trades to be checked by a get-together of deceitful performers. It gives a passed on, invariable, direct, protected, and auditable record. The blockchain can be directed straightforwardly and totally, allowing permission to all trades that have occurred since the essential trade of the system, additionally, can be checked and analyzed by any substance at whatever point [16, 23].

As the blockchain show gathering's information in a shackle of squares, though each square involves a lot of Bitcoin trades executed at a given time. Squares are associated and coordinated by a direction to the past square, outlining a shackle. To help and work by methods for the blockchain, network landed nobility need to give, the going with value: coordinating, limit, wallet organizations and mining [6]. As demonstrated by the limits they give, different sorts of center points can be fundamental for the association. The coordinating limit is imperative to check out the P2P association, this consolidates trade and square inciting. The limit work is subject for charge of a copy chain in the center point (the entire shackle for involved centers, and simply a piece of it for small centers). Wallet organizations give security keys that grant customers to mastermind trades, i.e., to work with their bitcoins. Finally, the digging work is obligated for making new squares by tending to the proof of work [5, 23]. The centers that play out the proof of work (or mining) are known as backhoes, and they get as of late made bitcoins, and costs, as a prize. The possibility of confirmation of work is one of the keys to engage thrustless arrangement in blockchain network [10, 24]. The check of work includes a computationally raised assignment that is fundamental for the time of squares. This work ought to be puzzling to address likewise, at the same time viably irrefutable once wrapped up. At the point when a digger completes the affirmation of work, it conveys the new square in the association and the rest of the association checks its authenticity before adding it to the chain. Since the period of squares is finished at the same time in the association, the square chain may unexpectedly fork in different branches (conveyed by different earthmovers). This mistake is settled by contemplating that the longest piece of squares is the one that will be considered as authentic. This, alongside the concentrated thought of the square age measure gives a novel, passed on thrustless-understanding segment.

It is computationally excessive for a malicious assailant to change a square and savage the square chain since the remainder of the accepted diggers would beat the attacker in the square age measure; therefore, the acknowledged piece of squares will ruin the one made by the aggressor. In explicit terms, all together for a controlled square to be effectively added to the chain, it is fundamental for handle the evidence of work speedier than the remainder of the affiliation, which is computationally pre-posterously over the top – it requires having control in any event of 51% of the planning assets in the affiliation. Taking into account the colossal computational limit expected to change the blockchain, the defilement of its squares is essentially anomalous. This recommends that, regardless of whether the people are not totally authentic about the use of Bitcoin, an arrangement is consistently reached in the association up to a huge portion of the affiliation is sketched out by certified people. The strategy proposed by Nakamoto was an extraordinary fomentation in the endurance of clashing entertainers in decentralized frameworks.

More encounters concerning the blockchain planning can be found in [5, 7]. Blockchain has besides given an improvement where the chance of brisk arrangement can be emerged. Right when everything is said in done terms, a sharp understanding intimates the PC programs or endeavors that award an agreement to be regularly executed/completed considering a great deal of predefined conditions. For instance, sharp game plans portray the application thinking that will be executed at whatever point an exchange happens in the trading of modernized money. In unbelievable plans, cut-off points and conditions can be portrayed past the trading of electronic financial structures, for example, the underwriting of resources in a particular degree of exchanges with non-money related parts, which makes it an ideal segment to create blockchain headway to various zones. Ethereum [8] was one of the pioneers blockchains to join astute arrangements. Today sharp courses of action have been participated in an enormous segment of existing blockchain use, for example, Hyperledger [9], a blockchain got ready for affiliations that awards pieces to be sent by the necessities of clients (savvy plans, associations or get-togethers among others) with the help of titanic relationship, for example, IBM, JP Morgan, Intel, and BBVA. The total of this has added to the progression of blockchain improvement to a colossal number of areas where the highlights offered by this advancement are required: loyal quality, lastingness, and auditability. Truth be told, blockchain is right now one of the top examination subjects of consistent occasions, with more than 1.4 billion dollars contributed by new associations alone in the hidden 9 months of 2016 as indicated by PwC [10]. Notwithstanding the way that the basic considered blockchain is immediate, its execution tends to a remarkable number of difficulties. This part presents the standard ones that its use achieves.

11.3 STORAGE CAPACITY AND SCALABILITY

Cut-off limit and adaptability have been essentially attended to in blockchain. In this headway, the chain is determinedly making, at a speed of 1 MB per block each 10 minutes in Bitcoin, and there are duplicates dealt with among the focuses in the affiliation. In any case, full focuses (a middle point that can absolutely underwrite exchanges and squares) store the full chain, aggregating necessities are massive. As the size makes, focuses require a continually extending number of assets, along these lines diminishing the design's ability scale. Furthermore, a greater than ordinary chain oppositely influences execution, for example, it expands synchronization time for new clients. Exchange support is a fundamental piece of the appropriated understanding show as focuses in the blockchain network are relied on to insist each exchange of each square. The measure of exchanges a square and the time between blocks, balance the computational force required and this clearly impacts exchange affirmation times. Thus, the arrangement show immediately influences the flexibility of blockchain networks.

Considering the trust model of Bitcoin and its flexibility objectives, Bitcoin-NG [11] proposes another Byzantine-deficiency liberal blockchain program which improves the course of action latency concerning Bitcoin. Litecoin [12] is truly muddled from Bitcoin, yet joins speedier exchange declaration times in addition, improved cut-off effectiveness taking into account the diminishing of the square age time and the confirmation of work, which depends upon Scrypt, a memory certified secret key-based key acknowledgment work. Phantom [13] is needed to improve the adaptability of Bitcoin by changing its chain choice principle. Off-chain game plans [14] are proposed to perform compromises the chain, developing the trade speed simultaneously as it amasses the likelihood of losing information.

Another proposal recommends lessening the spread delay [15] in the Bitcoin program; at any rate it can bargain the security of the affiliation. Instead of developing the adaptability on blockchain, BigchainDB [16] adds blockchain qualities to a critical information appropriated information base. BigchainDB consolidates the high throughput moreover, low inaction credits of monster information appropriated educational assortments with the consistent nature and decentralized strategy of blockchain. Another basic progress is the Inter Planetary File System (IPFS) [17]. IPFS is a program expected to store decentralized and shared files empowering a P2P appropriated record construction to make the web more secure, speedier, and more open. IPFS is planned to broaden the ampleness of the web at the same time as it takes out duplication similarly, tracks translation history for each record.

11.4   SECURITY (WEAKNESSES AND THREATS)

The Bitcoin program has experienced and been thoroughly isolated [18], and different weaknesses and security dangers have been found. The most eminent assault is the 51% or greater part assault [19]. This assault can happen if a blockchain part can manage over 51% of the mining power. In the current condition he/she can deal with the game plan in the affiliation. The effect and smart headway of mining pools (with GHash.io4 by chance appearing at 51% of the Bitcoin mining power in 2014), has expanded the likelihood of this assault occurring, which subsequently could bargain the reliability of Bitcoin. In addition, the producers in [20] talk about the chance of appearing at a ton of mining power through outcome. The display mining convincing force or P2P mining would help decline this issue. Different other agreement portions proposed for blockchains are in like way presented to greater part assaults, particularly those that concentrate the course of action among a predestined number of clients. The twofold spend assault remembers for spending an equivalent coin twice [21].

In Bitcoin an exchange ought to be viewed as stated simply after the square where the exchange is dealt with has a particular importance in the blockchain, routinely 5 or 6. This takes between 20 in addition, 40 min on normal [22]. There is a huge change in the certificate time since it relies on different sections. In quick bit conditions the transporter can't bear the cost of this relief. Properly, in these conditions, twofold spend assaults are up until this point conceivable. Fundamentally, race assaults can work in these conditions. To do this assault the client sends an exchange unmistakably to the seller, who perceives the exchange extremely brisk. By then the client sends different clashing exchanges to the affiliation moving the coins of the bit to himself. The following exchange is more disposed to be affirmed, and the merchant is cheated.

Fundamentally, the Finney [23] assault is a more current twofold spend, since it requires the interest of a digger. The extraordinary assaults of denial of service (DoS), man in the middle (MitM), or Sybil can besides debilitate the affiliation activity. Most P2P programs and IoT foundations are frail against such assaults, since they unflinchingly depend upon correspondences. In the overwhelm assault [24], aggressors can eat up a middle's affiliations, secluding it from the remainder of the affiliation and changing the perspective on the association for this middle. Code revives and streamlining in blockchain networks are usually kept up by part of the high-level money area are expected to improve their covered programs. These enhancements are known as touchy and hard forks in blockchain communicating. From one perspective, touchy fork give an update of the programming show that sees in reverse closeness with the past squares. This requires a reviving of most of the diggers to the new programming. Regardless, revived support can also be pardoned by most of the focuses keeping to the old principles. Obviously, hard forks convey a ludicrous change to the program, with no similarity with past squares and exchanges. Thus, all the focuses need to move to the most recent update, and focuses with more arranged variants will as of now not be perceived. The general public can be secluded when a hard fork occurs, accomplishing two indisputable forks of the affiliation. Hard forks can in like way be dropped on the off chance that they have not produced adequate agreement like SegWit2x [25]. Noted instances of that division are Ethereum and Ethereum Classic; and Bitcoin, Bitcoin Cash, and Bitcoin Gold.

What have of late been referred to as hard forks have advanced at the same time as the principal affiliations and these days they are doing combat with each other. Focuses and clients need to pick a translation, and the fork congruity will rely on these choices. Henceforth forks, particularly hard ones, can separate the local two totally exceptional blockchains and this can deliver a risk to the blockchain clients. An ordinary issue of virtual cash related standards, past the conversation consolidating their valid worth, is the issue of coin calamity. In the event that the wallet key fails to remember, there is no system to work with these mint pieces. It has been assessed that 30% of bitcoins are lost. At last, quantum selecting could be viewed as a danger to Bitcoin, since the figuring force of these PCs could break the security of electronic engravings. Also, improvement impels after some time and dependably new bugs and security enters are found. These updates and bugs can settle public blockchains with encoded information since blockchain information is consistent.

11.5   ANONYMITY AND DATA SECURITY

Security isn't realized in the Bitcoin program by plan. A basic element of Bitcoin is its straightforwardness. In blockchain each exchange can be checked, evaluated, and followed from the construction's altogether first exchange. This is undoubtedly an awesome new degree of straightforwardness that obviously assists with building trust. Anyway, this straightforwardness on influences security, paying little heed to the route that there is no snappy relationship among wallets and people, client absence of definition radiates an impression of being undermined despite the instruments that Bitcoin gives, for example, pseudonymous and the utilization of different wallets. In this sense, some exertion has been made to give more grounded secret recalls for Bitcoin. Obviously open virtual cash related standards, yet different applications wards on open blockchain advancement require a more brought level of security up in the chain, unequivocally those that regulate delicate information. Standard endeavors to manage the absence of definition issue in Bitcoin are Zero cash [26] and Zero coin [27] which propose that Bitcoin increments have totally dark exchanges, stowing interminably the sender, the beneficiary, and the real data. Monero [28] occupations a ring of engravings to make exchanges untraceable, so they can't be effortlessly followed back to some discretionary individual or PC.

Fundamentally, exchange blending associations or tumblers, given by Bitcoin Fog [29] and Bit Laundry can gather the secret. These associations separate exchanges into more unassuming segments and plan them to tangle exchanges for an expense. Notwithstanding, such associations ought to be skewed to robbery. In like manner, the coin-blending approach, from the start proposed in CoinJoin [30] serves to anonymize Bitcoin. The reasoning is that clients concede to joint segments, so it can as of now not be recognized that exchange inputs are from a similar wallet. Anyway, the past arrangement needed between clients, ordinarily performed by blending workers, could not have the important namelessness relying on the execution. Along these lines, this procedure convinced Dark Wallet [31], a program module that awards totally private astounding Bitcoin exchanges; Dash [32], known as the standard progressed money zeroed in on cloudiness and security; Mix Coin [33], that adds cryptographic obligation structures in addition, randomized blending charges to develop security; Coin Shuffle [34] that proposes a difference in CoinJoin to expand security; Coin Swap [35] that proposes a four-exchanges system considering the union of judges that get coins and make the part with disconnects coins; and Blind coin [36] that develops the secret of the blending worker. In light of everything, these endeavors to broaden secret in Bitcoin normally handle the possibility of keeping a liberated Bitcoin, and are in this way by and large blamed for empowering unlawful exercises, for example, the getting of unlawful things on the Darknet or assessment evasion. To build security, information in the blockchain can be encoded. Fowl of prey [37] stores blended exchanges. The Hawk compiler is in danger of translating the customary code made by engineers into cryptographic local people that empower data secret in exchanges. The Enigma project [38], being developed to encryption, parts information into unrecognizable bunches and dissipates them through the relationship, with the ultimate objective that no middle really pushes toward the information. It utilizes a decentralized off-chain passed on hash-table (DHT) available through the blockchain to store information references.

The issue of security in private blockchains can be managed in an astonishing way, since by definition they should give check, underwriting instruments. Regardless, even inside a private blockchain, people need to guarantee the security of their information. Predominant part [39] for example, is a private permissioned blockchain considering Ethereum that utilizes cryptography to restrict the perceivable nature of interesting information and division to build security of information. Multichain [40] masterminds client consents to bind perceptible quality and to present powers over which exchanges are permitted and which clients can mine. Rock chain [41] is correspondingly settled on Ethereum in addition, follows an information driven system, where public checks can be performed on private information, total outcomes can be acquired protecting information security. This methodology offers a gave report framework that licenses clients to direct information affirmation through sharp arrangements in Ethereum. Hyperledger Fabric [9] gives a streamed and versatile record zeroed in on enormous business conditions.

To give blockchain networks protection control, Hyperledger Texture gives a character control association and access control records through private channels where clients can manage and limit the enlistment to their shared data in the affiliation. Taking into account this fragment, individuals from the affiliation know each other through their public characters, in any case they don't need to know the data that it is partaken in the affiliation. Another way to deal with oversee tackle information security is to store delicate information outside the chain, regularly implied as the off-chain strategy [42]. Such a game plan favors structures that mediate a huge amount of information, since it is stunning to store them inside the blockchain. Additionally, they are especially appropriate for frameworks that manage remarkably precarious information that ought to have an all the closer access control, for example, clinical thought applications. Hence the public blockchain can be utilized to store anchor information, with the objective that affirmation to check the uprightness and time stamps of information is open. Clients can certify information without depending upon prepared experts, just by checking the blockchain, and information are securely dealt with outside. Evidently, these off-fasten sources should be deformity liberal and ought not present bottlenecks or single inspirations driving dissatisfaction. In [43] the creators propose utilizing a Kademlia, a striking DHT to store key respect sets (client characters and endorsements) to access, control, and breaking point information. In [44] a pointer and a hash to support the information acquired are dealt with in the chain. Along these lines, the information in the chain are joins to the private information, and the hash is the structure that confirms that the data acquired has not been changed. Access control structures for off-secure sources are given to guarantee that so to talk supported get-togethers can get to the data. The data would thus be able to be secured from outside sources in an ensured and checked manner with blockchain.

11.6 SMART CONTRACTS

In 1993, Nick Szabo depicted the sharp course of action as "An electronic expo that executes the game plans of an agreement." One of the essential highlights of a keen arrangement is that it has a way to deal with favor or self-execute authoritative enunciations. Until the rising of blockchain advancement, this was definitely unviable. Blockchain has ended up being the ideal headway to help sharp courses of action. Also, sharp plans have contributed fundamentally to the energy of blockchain, moreover this coupling has incited a second time of blockchains, all things considered known as Blockchain 2.0. The mix of regularly executed arrangements in a confided in climate without bound together control promises to change the manner by which current business is finished. Fundamentally, the cunning understanding code is dealt with on the blockchain, in like manner, each plan is perceived by a striking region, and for clients to work with it, they simply send an exchange to this region.

The right execution of the agreement is realized by the blockchain game plan show. Talented game plans present a ton of focal centers, for instance, cost decay, speed, precision, ability, and straightforwardness that have built up the presence of different new applications in a wide assortment of locales. Notwithstanding the way that Bitcoin offers a basic scripting language, it has ended up being deficient concerning, which has instigated the rising of new blockchain stages with encouraged awe-inspiring understanding support. The clearest savvy arrangement blockchain stage is Ethereum [8]. Ethereum is a blockchain with a perceived Turing complete programming language, that permits the meaning of sharp plans and decentralized applications. The code in Ethereum's game plans is written in "Ethereum virtual machine code," a low-level, stack-based bytecode language. Once in a while, monetary sharp courses of action foresee that enlistment should inform about obvious states and occasions. This information is given by the alleged prophets. These parts are fundamental for the gainful circuit of shrewd arrangements inside this current reality, yet endorsement, security and trust in prophets ought to be given [45].

The expected increases of smart plans don't come without cost, as they are powerless against a development of assaults [46–48] that bring new invigorating difficulties. Doling out agreement execution to PCs passes on with it two or three issues, since it makes them helpless against explicit issues, for example, hacking, bugs, defilements, or correspondence dissatisfactions. Bugs in arrangement coding are particularly central as a result of the irreversibly and unchanging nature of the design. Systems to check and ensure the right development of sharp game plans are critical for them to be overall and security embraced by customers also, suppliers. The standard support of the course of action thinking, and its accuracy are research regions where duties are should have been made in the years to come [49]. Essentially, ensured courses of action traditionally have plans or conditions that are not quantifiable. In this sense, there is still a great deal of work to be done to show the states of the game plans in sharp courses of action, with the target that they are representable and quantifiable for a machine to execute them. Furthermore, attempts to give contraptions for clients to have the decision to choose and understand savvy courses of action are required [50].

11.7 LEGAL ISSUES

The nonappearance of a focal position, the non-existent venturing part, and along these lines the complete inadequacy of oversight in Bitcoin is a drawing in and at the same time hazardous oddity. Bitcoin clients are regularly charged for utilizing the relationship for overwhelming purposes, and in this way the headway is connected with advancing or enabling illicit lead. Bitcoin, as the boss decentralized cryptographic money has made a ton of challenge [51]. From one point of view, concerning its worth, a few specialists guarantee it is a contortion [52] and that it will totally fall [53], while simultaneously others measure that its worth will appear at 100.000 dollars in 10 years [54]. The European Central Bank has given up its potential as a monetary progress [55]. In any case, concerning the nonattendance of association, different nations are growing new laws, endeavoring to organize the utilization of virtual cash-related standards (a guide of Bitcoin rule status is open at [56]).

The current condition has many shortcomings, and is clearly the explanation for its new fall [57]. Real results regarding cash-related designs are a gigantic worry, as they can plainly and conflictingly sway blockchain applications dependent on that cash. Different private and permissioned blockchain applications have definitely arisen. These are blockchains that award structures' endorsements to a predefined accomplice or set of colleagues. This can pass on several central focuses to support and insistence parts – for example, key recuperation or exchange recovery, and can in similar manner add to updating the issue of security and to reducing exchange latency. Truth be told, charming business zone openings have emerged in affirmation consolidation for bitcoins that would before long not be critical if the specialists managed this responsibility. The risk of mining pools controlling the relationship, close by different inadequacies, favors the movement of such blockchains, and governments are indisputably animated by an administered and controlled use of this improvement in different applications. In any case, this deduces that the thrustless affiliation will lose the faith to an untouchable trust affiliation, losing part of its encapsulation.

Moreover, this can make bottlenecks if arrangements unite concentrated parts. The highlights of these blockchains are nearer to the highlights of appropriated information bases. Obviously, the best way to deal with broadening trust in this headway could be the responsibility of governments or conceivably monstrous consortiums of relationship in their unexpected turn of events. Propelling activities toward this way will be useful [58, 59]. As of now, particular data is dissipated among various segments: government, schools, affiliations, etc. The data is spread across different segments and getting to it is redundant, notwithstanding, when said segments answer to a similar power, e.g., the public position. This prompt gets in getting to data regardless of a nonattendance of a dependable assistance that ensures the data. A reliable and by and large character association with the data of every individual would be badly arranged right now. Considering everything, every nation has its own laws and rules. Activities, for example, Alastria [60] hope to recollect various substances for the movement of a public facilitated blockchain, from public clerks to schools and selective associations.

They hope to draw in a public and legitimate wallet for every individual. Affiliations can also be significant for the affiliation. All things considered, each precious wallet could be an electronic verification of assets, affiliations where he/she has worked, higher directions, etc. This data could be utilized in a certifiable, strong route for specific associations. For example, due to an arranged agent gathering, competitors could share their higher trainings and work experience data with the inspectors. This data is solid and obvious by definition. These activities are the best way to deal with extending blockchain inside government establishments, additionally, the hidden stage in making an ordinary and administrative development for blockchain structures. Moreover, this could in like way uphold the authentic exchanges of everybody to get their data, the information moves between nations, the diminishing of savage data and an anticipated trade-off between individuals, affiliations, government, and colleges. Notwithstanding, this likewise achieves a fundamental procedure to acquire essentially private data, so the protection and security considered in the remainder of the paper ought to on a fundamental level go unclearly with these activities.

11.8　CONSENSUS

Understanding fragments [61–63] are subject for the fairness of the data contained in blockchain, while making sure about against twofold spend assaults, and consequently are a central piece of blockchain improvement. The last objective is to accomplish understanding in a dispersed relationship without focal topic specialists and with people who don't really trust one another. The arrangement subject to the check of work (PoW), which has worked so effectively in Bitcoin, powers tractors to deal with a computationally-centered sufficiently certain errand to make another square. Exactly when kept an eye on, the game plan is circled and the new square is added to the chain. The new square is spread across the association and the remainder of the people check it and append it to its nearby by blockchain duplicate. This cycle can simultaneously happen in various pieces of the affiliation. This is the clarification the chain is in all honesty a tree. There are a couple, huge branches existing at the same time in the blockchain network. Right when partners add another square, they likewise need to watch that the branch is the one with the most accumulated work (burden), that is, the longest chain which is recognized to be the genuine one. These awards agree to be refined rapidly. A key disadvantage is that PoW makes Bitcoin subject to energy use. The as of late referred to 51% assault is a typical assault on the Bitcoin program.

Similarly, the prodding powers in PoW are incredibly pushing centralization as the advancement of mining pools confirm. This close by the mint lessening, reward reduction and charge expansion could bargain the design [64] later on. Clearly, PoW has certain burdens, for example, high inaction, low exchange rates, and high energy use that makes it unsatisfactory for specific applications. As conveyed, the idleness, or of course block rehash of 10 min may also be unfeasible in different conditions. Ignoring this, a few phases do utilize or have changed PoW, for example, NameCoin, LiteCoin, Ethereum, DogeCoin, and Monero. Primecoin, for example, mitigates the energy occurrence by proposing critical computationally raised assignments – for example, the indivisible numbers search that can have applications close by. Clearly, different endeavors to change PoW have really been proposed, likely criticizing the unusualness that this change suggests, everything considered it isn't certain about the slim chance that they reveal the security properties similarly as PoW. The most standard elective way to deal with oversee understanding in blockchain is the Proof of Stake (PoS). It depends upon the way that those clients who own more coins, are faster on the persistence and the right working of the construction, and in this way are the most reasonable to pass on the commitment of ensuring the framework. Essentially, the thought behind utilizing PoS is to move the chance expenses from outside the framework to inside the design.

The figuring erratically picks the client liable for the course of action of each square, thinking about the number of coins he/she has. A normal report is that this method doesn't offer motivations to focuses to pick the right square (known as the nothing being alluded to issue). In addition, it is negative as in it impels overhaul of the rich. PoS was from the start utilized by Peercoin and later in Nextcoin, Nxt [65], Pine for and Ethereum. An assortment of PoS is the Delegate PoS (DPoS) of BitShares, Monax, Lisk, or Tendermint. In BitShares [66], different picked onlookers favor checks and time stamps of exchanges by reviewing that them for blocks. The political race is performed by projecting a democratic structure, and each time an onlooker sufficiently makes a square it is reviewed. This methodology licenses experts to set the square inactivity, block evaluate and demand exchanges a second.

The Leased Proof of Stake (LPoS) licenses clients to rent stores to different focuses, with the target that they will without a doubt be picked for block creation, broadening the measure of electable people, and accordingly diminishing the likelihood of the affiliation being compelled by a solitary get-together of focuses. Prizes are for the most part shared. The Proof of Burn (PoB) [67] proposes eating up coins, that is, sending them to an obvious unspendable territory, to flow another square. Like PoW, PoB, is difficult to do and simple to assert, regardless, oddly requires no energy utilization. Additionally, PoB has some money related outcomes that add to a steadier environment. Nem's [68] way of dealing with overseen course of action, called Proof of Importance (PoI), relates a significance respect with each record, in this way gathering a standing construction in the affiliation. The possibility of being picked to make a square relies on this worth, its tally likewise considers the measure of coins and the measure of exchanges performed. With everything considered, profitable association action is in addition redressed, not simply the aggregate, driving the obliging conduct of the clients.

Other broadened assortments are the action test (PoA), a cream approach that combines both PoW and PoS, and the Elapsed Time Test (Artist) made by IBM that utilizes an emotional decision of boss to mine each square dependent on runtimes inside solid execution conditions. The Proof of Capacity (PoC) [69], regardless called confirmation of breaking point or space, utilizes open hard drive space as opposed to figuring assets. This methodology is utilized in Permacoin, SpaceMint also, Burstcoin. Private blockchains have express highlights, since the amount of people is all around lower than public blockchains and are semi-solid. They are all things considered took on the design with a predefined set of endorsements. These frameworks, subsequently, require express arrangement portions that fit these attributes. A section of these elective parts are Paxos [70], made by Lamport and Microsoft dependent on state machine replication; Chubby [71], considering the past and made by Google, which is depicted as an orbited obstructing help.

These strategies have the piece of slack that they are assortments of formal tallies, and in this way their highlights have been definitively shown. Boat [72] which detaches key pieces of comprehension, for instance, selection of pioneers, record replication and security, and powers a more basic level of consistency to lessen the measure of states that should be thought of. The Practical Byzantine Fault Tolerance (PBFT) assessment, which depends upon state machine replication in addition, reproduce administering for admission to state change, is utilized in Hyperledger and Multichain. Sifter [73], treats the blockchain as an exposure, executing endeavors and separating the yield of each copy. In the event that there are divergences between the ages, the development isn't supported. Another assortment of the PBFT is the Byzantine game plan Federated Byzantine Agreement (FBA). In FBA each part keeps a synopsis of confided in people and monitors things for these people to respect an exchange prior to being considered exchanged. It is utilized in Ripple [74]. Famous [75] is another assortment that utilizes the lion's offer and halfway bigger part thought. The bigger part is a ton of focuses, enough to agree, the halfway lion's offer is a subset of a larger part with the capacity to persuade another given community about the strategy. HDAC [76] is a framework, as of now being executed, which proposes an IoT Contract and M2M Transaction Stage dependent on Multichain.

HDAC is remarkably exceptionally intended to IoT conditions. It utilizes the ePow plan calculation whose principal targets are to move the assistance of different mining habitats in addition, to upset silly energy squander. At long last, the HC Consensus, proposed in Hydrachain, depends upon a quick overview of validators of which close to 33% are beguiling. To summarize, course of action systems in open blockchains have been completely proposed now deficiently, officially represented. It is required to comprehend the ensures they offer and their weaknesses going prior to utilizing them [77]. In this sense, the examination neighborhood with the business needs to seek after the support of these instruments to show their validity. To the backwards, private blockchains have acknowledged legitimate prominent courses of action, yet the restricted synopsis of people in these blockchains likewise limit the variety and capacity of businesses.

11.9   IOT AND BLOCKCHAIN INTEGRATION

The IoT is changing and refreshing manual cycles to make them a piece of the general time, getting volumes of information that gives information at unimaginable levels. This information is engaging the improvement of savvy applications, for example, the improvement of the association and the individual satisfaction of inhabitants through the digitization of associations in the metropolitan organizations. Over the most recent couple of years, flowed figuring movements have added to giving the IoT the critical comfort to investigate and evaluate data and change it into consistent activities and information [1]. This earth-shattering progression in the IoT has opened up new area, for example, systems to get to and share data. The open information viewpoint is the pioneer in these activities. Regardless, perhaps the essential inadequacies of these activities, as has happened in different conditions, is the nonattendance of confirmation. Joined models like the one utilized in cloud getting ready have commonly added to the movement of IoT. Notwithstanding, seeing information straightforwardness they go likely as secret segments and network people don't have an idea of where and how the data they give will be utilized.

The coordination of promising advances like IoT and cloud planning has been shown to be immense. Similarly, we see the best limit of blockchain in improving the IoT. Blockchain can drive the IoT by offering a confidence in sharing assistance, where data is solid and can be obvious. Information sources can be seen at whatever point and information stays enduring over the long haul, expanding its security. In the conditions where the IoT data ought to be safely part between different people this mix would address a key insubordination. For example, a concentrated detectable quality in different food things is an essential point of view to guarantee cleansing. Food prominence could require the responsibility of different people: making, managing, treatment, stream, etc. An information spill in any piece of the chain could provoke double dealing and upset the examples of the excursion for corrupting which can genuinely affect occupant's lives and cause giant money related expenses to affiliations, zones and nations in view of a foodborne scene [78]. An unmatched control in these districts would amass food success, improving the information parting among people and decreasing the pursuit time in view of a foodborne scene, which can save human lives. Likewise, in different zones, for example, sharp

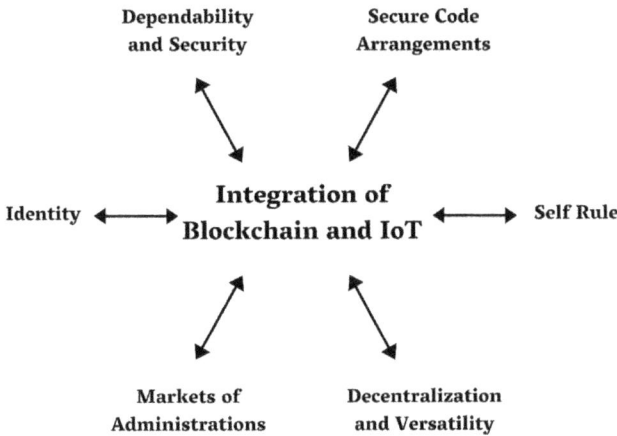

FIGURE 11.1 Blockchain – IoT Integration.

metropolitan organizations besides, sharp vehicles, sharing solid information could maintain the prospect of new people in the characteristic systems and add to improve their associations and their get-together. Accordingly, the utilization of blockchain can upgrade the IoT with reliable and secure data. This has begun to be viewed as alluded to in [79], where blockchain progression is seen as the best way to deal with address adaptability, security, and unwavering quality issues identified with the IoT viewpoint.

From our perspective IoT can enormously profit by the accommodation given by blockchain and will assist with advancing making current IoT drives. It is critical that there are so far an incomprehensible number of examination inconveniences and open issues that should be concentrated to faultlessly utilize these two improvements together and this examination subject is so far in a starter stage (Figure 11.1).

Considerably more explicitly, enhancements that this joining can bring unite (at any rate are not restricted to):

i. **Decentralization and versatility**: the move from a solidified planning to a P2P dissipated one will forgo key issues of dissatisfactions and bottlenecks [80]. It will also help forestall conditions where a few astounding affiliations control the managing and cut-off of the data of countless individuals. Different inclinations that go with the decentralization of the arrangement are an improvement of the variety to internal frustration and framework adaptability. It would reduce the IoT storerooms, and moreover add to improving the IoT flexibility.

ii. **Identity**: utilizing a regular blockchain framework part can see each and every gadget. Information gave and managed into the framework is constant and extraordinarily sees authentic information that was given by a gadget. In addition, blockchain can give confidence in appropriate endorsement and underwriting of contraptions for IoT applications [81]. This would address an improvement in the IoT field and its people.

iii. **Self-rule**: blockchain headway interfaces with front-line application high-lights, making conceivable the improvement of wonderful autonomous resources and equipment as an association [82, 83]. With blockchain, gad-gets are useful for collaborate with each other without the thought of any workers. IoT applications could profit by this supportiveness to give con-traption skeptic and decoupled-applications.

iv. **Dependability**: IoT data can stay steady and appropriated over the long haul in blockchain [84]. People from the framework are ready to attest the realness of the information and have the insistence that they have not been altered with. In like manner, the improvement connects with sensor infor-mation conspicuousness and commitment. Unwavering quality is the funda-mental piece of the blockchain to get the IoT.

v. **Security**: data and exchanges can be guaranteed about on the off chance that they are dealt with as exchanges of the blockchain [85]. Blockchain can treat contraption message trades as exchanges, embraced by magnificent game plans, in this route guaranteeing about exchanges between gadgets. Current secure standard protocols utilized in the IoT can be improved with the utilization of blockchain [86].

vi. **Market of Administrations**: blockchain can breathe life into the formation of an IoT environment of associations and information business centers, where exchanges between peers are conceivable without topic specialists. Microservices can be effectively passed on and more modest than typical seg-ments can be security made in a trust less climate [87–89]. It would improve IoT interconnection and the passageway of IoT information in blockchain.

vii. **Secure Code Arrangement**: manhandling blockchain secure-constant breaking point, code can be flourishing and safely collided with contrap-tions [80, 90]. Producers can follow state plus, empowers with the most raised sureness [85]. IoT middleware's can utilize this comfort to safely fortify IoT gadgets.

Another point of view to consider is identified with the IoT affiliations, i.e., the cor-respondence between the fundamental IoT foundation. While arranging blockchain, it should be picked where these affiliations will occur: inside the IoT, a mutt configu-ration including IoT and blockchain, or through blockchain. Haze planning [91] has besides upset the IoT with the combination of another layer between appropriated preparing and IoT gadgets and could also engage this joining.

11.10 CHALLENGES IN BLOCKCHAIN – IOT INTEGRATION

This part accepts the fundamental inconveniences to be watched out for while applying blockchain progression to the IoT locale. The coordination of blockchain progression with the IoT isn't minor. Blockchain was anticipated an internet condition with daz-zling PCs, also, this is a long way from the IoT reality. Blockchain exchanges are intentionally stepped, and hence contraptions outfitted for working with the money should be furnished with this accommodation. Joining blockchain into the IoT is attempting. A portion of the apparent troubles are introduced in this piece (Figure 11.2).

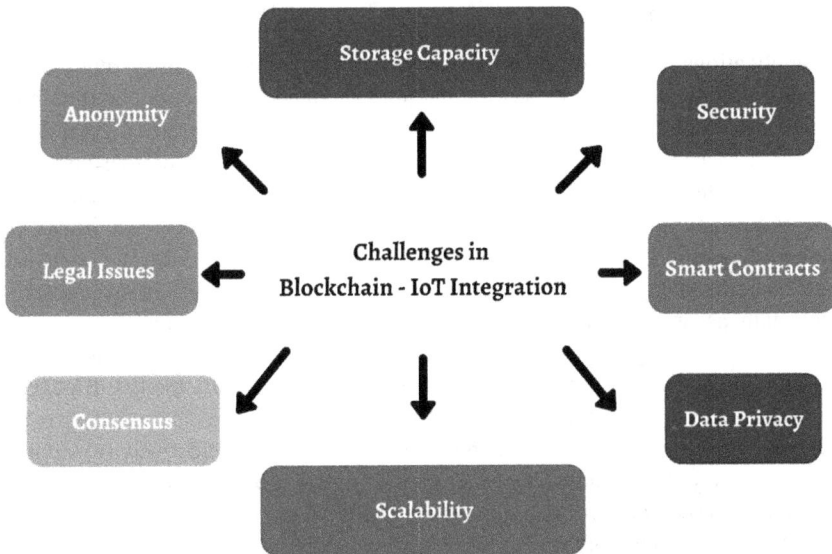

FIGURE 11.2   Challenges in Blockchain – IoT Integration.

11.10.1   STORAGE CAPACITY AND SCALABILITY

As imparted, storing up limit and adaptability of blockchain are so far being discussed, yet regarding IoT applications beyond what many would consider possible and adaptability objectives make these inconveniences fundamentally more significant. In this sense, blockchain may show up, evidently, to be precluded for IoT applications, at any rate there are propensities by which these restrictions could be diminished or avoided in general. In the IoT, where contraptions can convey gigabytes (GBs) of information powerfully, this constraint delivers a remarkable deterrent to its mix with blockchain. It is perceived that some current blockchain use can basically cycle a couple of exchanges for reliably, so this could be a presumably bottleneck for the IoT. Also, blockchain isn't intended to store a huge load of information like those made in the IoT.

A trade-off of these advances should manage these troubles. As of now, a tremendous heap of IoT information are dealt with and essentially a limited part is helpful for eliminating information and making works out. In the forming various techniques to channel, standardize, and pack IoT information to lessen them have been proposed. The IoT fuses installed gadgets, correspondence and target associations (blockchain, cloud), in this way theory resources in the extent of information that the IoT gives can profit various layers. Information squeezing variable can help transmission, managing assignments and breaking point of the high volume of IoT information made. Normal practices don't as a rule require extra, head data, rather than specific information. To wrap things up, blockchain, and particularly its arrangement show which causes its bottleneck, could in like way be accustomed to expand the data transmission and diminishing the lethargy of its exchanges this way empowering a preferable change over the IoT showed up by the instance of Bitcoin-NG [11].

11.10.2   SECURITY

IoT applications need to supervise security issues at various levels, at any rate with an extra multifaceted plan because of the nonattendance of execution and high heterogeneity of contraptions. Moreover, the IoT situation incorporates a great deal of properties that sway security, for example, convenience, inaccessible correspondence, or scale. An escalated evaluation of security in IoT is past the level of this paper at any rate point by point studies can be found in [97–100]. The broadening number of assaults on IoT affiliations, and their bona fide impacts, make it amazingly more basic to make an IoT with more eccentric security. Different specialists see blockchain as an imperative improvement to give the really significant security upgrades in IoT. Regardless, one of the critical inconveniences in the joining of the IoT with blockchain is the reliability of the information made by the IoT. Blockchain can guarantee that information in the chain is consistent and can perceive their changes, everything considered when information shows up effectively ruined in the blockchain they stay dreadful. Degenerate IoT information can ascend out of different conditions secluded from noxious ones. The thriving of the IoT planning is influenced by different factors for example, the climate, people, mutilation, and the error of the contraptions. In some cases, the genuine contraptions and their sensors moreover, actuators dismissal to work fittingly from the most punctual beginning stage. The current condition can't be perceived until the contraption being insinuated has been endeavored, or on the other hand generally it winds up being appropriate for a long time and changes its lead without any justifiable cause (cut off, revamp obsolete quality, etc.). Despite these conditions, there are different dangers that can affect the IoT, for example, sneaking around, refusal of association or controlling [98]. Consequently, IoT gadgets ought to be all around endeavored before their coordination with blockchain and they ought to be found and exemplified in the ideal spot to stay away from genuine harm, in spite of including procedures to see gadget disappointments when they occur. These gadgets will without a doubt be hacked since their objectives limit the firmware empowers, protecting them from activating over potential bugs or security breaks.

Besides, it is once in a while hard to resuscitate contraptions independently, as in by and large IoT plans. Subsequently, run-time invigorating and reconfiguration portions ought to be set in the IoT to keep it pursuing some time. Activities, for example, GUITAR [101] and REMOWARE [102] draw in affiliation and firmware resuscitates in run time and are critical to guarantee an ensured breaker of the IoT with blockchain over the long haul. The IoT and blockchain coordination can in like way have repercussions on the IoT exchanges [86]. As of now, IoT application programs, for example, CoAP and MQTT utilize other security programs for example, TLS or DTLS to give secure trades. These protected programs are befuddling and significant in spite of requiring a unified association and association of key foundation, routinely with PKI.

In the blockchain network each IoT gadget would have its own GUID (Global Unique Identifier) and unequal key pair introduced once associated with the affiliation. This would improve current security programs which usually need to trade PKI bolsters and would permit them to be utilized in gadgets with lower limits. One recognizable IoT project with respect to security with a blockchain assurance is Filament [83]. Fiber is a stuff and programming strategy that offers accommodation to Bitcoin-based segments and awesome game plans in IoT. Fiber gadgets have installed cryptoprocessors that help five programs: Block name, Tele hash and gifted game plans to work, and besides Penny back and BitTorrent programs. The gadget character the heads is finished with block name, while Tele hash, an open-source execution of Kademlia DHT, gives secure blended trades, and unbelievable courses of action portray the manner in which a gadget can be utilized.

11.10.3 ANONYMITY AND DATA PRIVACY

Different IoT applications work with secret information, for example precisely when the contraption is related with an individual, for example, in the e-flourishing situation, it is fundamental to address the issue of information affirmation and absence of clearness. Blockchain is introduced as the ideal reaction for address character the bosses in IoT, regardless as in Bitcoin, there might be applications where absence of lucidity should be ensured. This is the situation of a wearable with the capacity to camouflage the personality of the individual when sending particular information, or breathtaking vehicles that shield the security of the plans of clients. The issue of information protection in immediate and public blockchains has as of late been talked about, close by a touch of the current strategies. Notwithstanding, the issue of information confirmation in IoT contraptions incorporates more trouble, since it begins at information assortment and releases up to the correspondences and application levels. Guaranteeing about the contraption with the target that information are dealt with safely and not got to by individuals without consent is a test since it requires the joining of security cryptographic programming into the gadget. These overhauls should consider the impediment of assets of the contraptions and the limitations identified with money related sensibility.

Different advancements have been utilized to guarantee about correspondences utilizing encryption (IPsec, SSL/TLS, DTLS). IoT gadget impediments occasionally make it vital for utilize less-obliged contraptions, for example, entries to join these security programs. The utilization of cryptographic equipment could restore cryptographic activities and dodges the over-weight of complex secure programming shows. Security of information and protection are key difficulties for IoT, utilizing blockchain improvement the issue of character the board in IoT can benefit from outside intervention. Trust is another essential segment of the IoT where the combination of blockchain can anticipate a work. In [103] the significance of trust in IoT structures is seen as one of the central objections to guarantee its thriving. Information reliability procedures are another choice to guarantee information access at the same time as they put forth an attempt not to over-inconvenience blockchain with the beast extent of information made by the IoT. This can accomplish open frameworks, in any case with a fit and confined permission control. MuR-DPA [104] gives dynamic information animates and fruitful check in any case open dissecting attestation.

In [105] the producers guarantee the information content through another security defending public investigating framework. For a wide survey of goodness assertion system, recommend [106]. To wrap things up, there are laws that regulate information security, for instance, the EU's information security orchestrates that should be re-examined to cover the new models that the improvement makes conceivable. The assurance of blockchain as a real stage should deliver these principles to guarantee information security keeping the law

11.10.4   SMART CONTRACTS

Smart contracts are programs that run on blockchain after fulfilling predefined conditions. A smart contract is an agreement between two people in the form of a computer code. They run on blockchain, so they are stored in a public database and cannot be changed. Magnificent plans have been seen as the executioner utilization of block-chain headway, yet as alluded to there are a few difficulties yet to be managed. IoT could profit by the utilization of watchful arrangements, regardless the way where they fit into IoT applications is different. From a common-sense perspective, a plan is a gathering of code (cut-off points) and information (conveys) that live in a particular blockchain address. Public cut-off points in an arrangement can be called by contraptions. Cut-off points can also fire occasions, applications can tune in for them in sales to appropriately respond to the occasion finished. To change the condition of the game plan, that is, to adjust the blockchain, an exchange ought to be spread in the affiliation. Exchanges are embraced by senders moreover, ought to be perceived by the affiliation. The IoT can perceive and provoke over the internet in different zones [1]. For example, In the food perceptibility model, food bundling would be outfitted with sensors with the capacity to check natural conditions and interface with the blockchain (sign exchanges). In the blockchain an agreement would offer capacities to begin dispatching, complete the pattern of transportation and log and question evaluations.

Constantly, execution of fast game plans is done in a solitary community anyway at the same time the code execution is finished by different focuses. This spread is basically refined for the underwriting measure, instead of utilizing it to pass on undertakings. The IoT has utilized the hovered furthest reaches of flowed preparing and gigantic information to develop its preparing power. Beginning now and for a significant length of time, information mining frameworks have had the choice to address the IoT information when everything is said in done, connecting with an unmatched comprehension of the IoT, i.e., the arranging power extended by coursed preparing. Colossal information has connected with the arranging of a huge load of information at the same time, permitting information to be disengaged from huge datasets, which was ahead of time extremely hard to do. In the mix of IoT with block-chain, magnificent courses of action ought to use their spread nature to draw in the dealing with limits gave in different ideal models (gigantic information and scattered enrolling) and required in the IoT. Marvelous courses of action ought to comparably consider the heterogeneity similarly, necessities present in the IoT. Segregating and collecting instruments ought to be upgraded by astonishing courses of action to empower applications to address the IoT relying on the specific circumstance and necessities. A divulgence instrument could connect with gadget solidification on the fly, making these applications considerably more excellent. Finally, incitation instruments obviously from sharp plans would draw in snappier responses with the IoT.

11.10.5 LEGAL ISSUES

The vision of an unregulated blockchain is key for its substance, furthermore, some-what liable for the accomplishment of Bitcoin. As seen, blockchain, unequivocally concerning virtual cash related constructions, has passed on with it an immense heap of discussion regarding realness. The need, or of course opportunity, to present con-trol sections over the affiliation has come as permissioned, private, and consortium blockchains. The IoT space is besides affected by a nation's laws or rules with respect to information security, for example the information insurance order. A large portion of these laws are getting away from date and ought to be changed, particularly since the climb of new perilous degrees of progress, for example, blockchain. The move-ment of new laws additionally, rules can energize the accreditation of security high-lights of contraptions, and in this way help make the most secure and trusted in IoT affiliation. In this sense, laws that supervise data confirmation plus, data managing are as of recently a critical test to be dealt with in IoT and will consequently be a widely more imperative test at whatever point utilized in mix in with blockchain.

As imparted, the nonappearance of rule makes wounds, since parts for private key recovery or reset, or exchange inversion are impossible. Some IoT applications imag-ine an around the planet, exceptional blockchain for contraptions, regardless it is indistinct if such an affiliation is proposed to be controlled by creators or open to clients. Regardless, it is normal it will require lawful principle. These standards will influence the inescapable fate of blockchain furthermore, IoT and as such may truly annoyed the decentralized and free nature of blockchain by presenting a controlling, joined part, for example, a country.

11.10.6 CONSENSUS

Concerning IoT applications, the limited asset nature of gadgets chooses them pre-cluded for partaking in game plan sections, for example, PoW, direct. As imparted, there are a wide course of action of recommendations for getting shows, dismissing how they are, if all else fails, youthful and have not been endeavored enough. Asset fundamentals rely on the specific sort of plan show in the blockchain network. Typically, approaches will when everything is said in done delegate these errands to portals, or some other unconstrained gadget, fit for giving this comfort. On the other hand, off-chain game plans, which move data outside the blockchain to lessen the high inertness in blockchain, could give the incentive notwithstanding the path that there are activities to merge blockchain full focuses into IoT contraptions [92, 93], mining is so far a fundamental test in the IoT because of its restrictions. IoT is mainly made out of resource constrained gadgets in any case all around the IoT has a possibly gigantic preparing power, considering that it is ordinary that the number of contrap-tions in it will show up at any place some spot in the extent of 20 and 50 billion by 2020. Examination attempts should zero in on this field and effect the appropriated nature and by and large capacity of the IoT to change the course of action in the IoT.

In Babelchain [107] a novel game plan show called Proof of Understanding (PoU) that means to change PoW for IoT applications is proposed. With less energy utiliza-tion, the show, assessing utilizing excavators to address hash puzzles, proposes deciphering from various shows. Hence the exertion is more based on critical assessment while at the same time managing a central matter of interest in IoT correspondences. Companions in the relationship as opposed to conceding to exchange status, concur on message significance (plan, content in like manner, development). Moreover, the blockchain information offers data to learn, similar to a learning set.

11.11   PLATFORMS AND APPLICATIONS

In reality blockchain stages and applications have risen up out of different masterminded zones, because of the central focuses that this advancement offers. This part looks at the most specialist applications and stages that blend the IoT and blockchain.

11.11.1   Blockchain Stages for IoT

Blockchain has been perceived as an interesting progression that can unequivocally affect different undertakings. The measure of stages is so high and in steady change that it is difficult to isolate them all, in this part we rotate around the most acclaimed and all things considered reasonable for IoT spaces. Bitcoin was the guideline electronic cash and the first blockchain stage. It gives a system to do trade related exchanges out a quick, inconspicuous, and solid way, which can be formed into applications as a secured partition structure. In IoT space, free gadgets can utilize bitcoins to perform downsized divides, working generally as wallets. In general, when the use of blockchain is kept to limited scope divides, applications are connected to the cash, which can be an impediment, since corrupting of the coin can oppositely affect the application. As imparted, utilizing sharp arrangements is an ordinary arrangement while arranging blockchain with IoT. Bitcoin wires a scripting language that awards express conditions to be set when completing exchanges. Regardless, the scripting is particularly restricted separated and other sharp arrangement stages. As alluded to one of the stages that has had a basic impact as of late is Ethereum [8].

Ethereum was one of the pioneer blockchains in including sharp plans. Ethereum can be depicted both as a blockchain with a trademark programming language (Solidity), and as a plan based virtual machine running all around the world (Ethereum Virtual Machine EVM). The joining of savvy arrangements moves the blockchain away from cash related rules moreover, underpins the blend of this improvement in new areas. This near to its dynamic and clearing neighborhood Ethereum the most standard stage for making applications. Most IoT applications use Ethereum or are sensible with it. The easiest methodology is to depict a sharp game plan where contraptions can pass on their measures and approaches that respond to changes.

Hyperledger [9] has besides had a striking effect. Hyperledger is an open-source stage on which different undertakings identified with blockchain have been made, among them Hyperledger Fabric, a blockchain denied of endorsements and without cryptographic money on which business utilize like IBM's blockchain stage are based. It gives various parts to course of action additionally, enrolment. Dissipated application can be made in the blockchain utilizing thoroughly accommodating dialects. IoT contraptions can supply information to the blockchain through the IBM Watson IoT Platform, which oversees gadgets and awards information evaluation and sifting. IBM's Bluemix stage supports the mix of blockchain headway by offering it as an association. The use of this stage speeds up application prototyping, and two or three usage cases have been made. There is a constant undertaking on food conspicuousness that utilizes this stage [108].

The Multichain stage permits the creation and plan of private blockchains. Multichain utilizes an API that expands the point of convergence of the essential Bitcoin API with new supportiveness, permitting the main gathering of portfolios, resources, consents, exchanges, and so forth. Also, it offers a solicitation line gadget for collaborating with the association, and various customers that can relate through JSON-RPC with the relationship, for example, Node.js, Java, C # and Ruby. Multichain is a fork of Bitcoin Core, its source code aggregates for 64 cycle structures. In [109] Multichain blockchain pack is passed on three place focuses, one of them an Arduino board, as a proof of thought about an IoT–blockchain application. Litecoin [12], as imparted, is truly undefined from Bitcoin, in any case consolidates quicker exchange affirmation times and improved putting away productivity by uprightness of the reducing of the square age time (from 10 min to 2.5) and the confirmation of work, which depends upon Scrypt, a memory certified secret word-based key acknowledgment work. This recommends that the computational necessities of LiteCoin focuses are lower, so it is more reasonable for IoT. Lisk [110] offers a blockchain stage in which sub blockchains or sidechains can be depicted with decentralized blockchain applications and a decision of cryptographic kinds of money to use (for example Bitcoin, Ethereum, and so on) Known as the blockchain stage for JavaScript designers, Lisk likewise offers backing to make and send decentralized applications inside the stage to be utilized obviously by end clients, making a characteristic game plan of interoperable blockchain associations. The applications made can utilize LSK money or can make custom tokens. Lisk utilizes Delegated confirmation of stake plan. Lisk is working with Chain of Things to inspect whether blockchain improvement can be appropriate in setting up security inside IoT.

Lion's offer [39] is a blockchain stage made to give the cash related associations industry with a permissioned execution of Ethereum with help for exchange and game plan security. It permits different course of action systems and accomplishes information security through cryptography and division. The stage has really combined Zero Cash headway to cloud all prominent data about an exchange. The Quorum stage has been utilized by Chronicled [111] to make secure relationship between genuine resources and blockchain. HDAC [76] is an IoT comprehension and M2M exchange stage considering blockchain right now a work in progression. The HDAC framework utilizes a mix of public and private blockchains, and quantum self-self-assured number age to guarantee about these exchanges. The HDAC modernized money drew in open blockchain can be sufficient utilized with various private blockchains. Hdac IoT contract evidence of thought will be dispatched for the current year [112].

11.12 CONCLUSION

Unsafe headways dependably make exceptional conversation. Despite the route that there are different killjoys of virtual monetary rules, it appears to be unquestionable that the improvement that maintains them is a tremendous mechanical vexation. Blockchain is making a plunge for the extended length. Regardless, adjusting the headway without sufficiently ensuring its development or applying it to conditions where the expense doesn't remunerate the improvement are chances into which one can fall with no issue. Hence, the potential gains of applying blockchain to the IoT ought to be explored carefully and treated with care. This chapter has given an assessment of the significant inconveniences that blockchain and IoT should address with a definitive target for them to feasibly partake. We have perceived the central issues where blockchain progression can help improve IoT applications. An assessment has moreover been given to show the good judgment of utilizing blockchain focuses on IoT gadgets. Existing stages and applications have also been explored to complete the evaluation, offering a flat-out outline of the joint effort between blockchain progression and the IoT point of view. It is standard that blockchain will change the IoT. The mix of these two progressions ought to be tended to, considering the difficulties perceived in this paper. The selection of rules is essential to the possibility of blockchain and the IoT as a portion of government frameworks. This decision would quicken the connection between occupants, governments, and affiliations.

Game plan will also expect an essential occupation in the possibility of the IoT as an element of the mining measures and spreading altogether more blockchains. Notwithstanding, a dualism between information conviction and enabling the wire of presented gadgets could emerge. Finally, past the adaptability and breaking point which sway the two turns of events, research attempts ought to comparatively be made to guarantee the security and protection of fundamental headways that the IoT and blockchain can transform into. One of the focal worries about blockchain, and particularly cryptographic kinds of money, stays in its unusualness which has likewise been mauled by individuals to mishandle the current condition. The coordination of the IoT and blockchain will generally develop the use of blockchain, to set up cutting edge sorts of money on a comparable level as current guard cash.

REFERENCES

1 M. Díaz, C. Martín, B. Rubio, State-of-the-art, challenges, and open issues in the integration of internet of things and cloud computing, *J. Netw. Comput. Appl.* 67 2016 99–117.

2 J. Rivera, R. van der Meulen, Forecast alert: Internet of things – Endpoints and associated services, worldwide, 2016, Gartner 2016.

3 *World Health Organization Food safety fact sheet*, 2017. Available online: http://www. who.int/mediacentre/factsheets/fs399/en/ (Accessed 1 February 2018).

4 *17 Blockchain disruptive use cases*, 2016. Available online: https://everisnext. com/2016/05/31/blockchain-disruptive-use-cases/ (Accessed 1 February 2018).

5 S. Nakamoto, *Bitcoin: A peer-to-peer electronic cash system*, 2008. Available online: https://bitcoin.org/bitcoin.pdf (Accessed 1 February 2018).

6 A. M. Antonopoulos, *Mastering Bitcoin: Unlocking Digital Cryptocurrencies*, O'Reilly Media, Inc., 2014.

7 K. P. Arjun et al., Distributed computing and/or distributed database systems, *Blockchain Platforms and Applications*, ISBN 9780367533403, Auerbach Publications, CRC Press, September 2020.

8   Dr. K. Saini, *A future's dominant technology Blockchain: Digital transformation*, in *IEEE International Conference on Computing, Power and Communication Technologies 2018 (GUCON 2018)* organized by Galgotias University, Greater Noida, 28–29 September, 2018. doi:10.1109/GUCON.2018.8675075.

9   S. Dutta, K. Saini, Securing data: A study on different transform domain techniques, *WSEAS Trans. Syst. Control* 16 2021. E-ISSN: 2224-2856. doi:10.37394/23203.2021.16.8

10 J. Kennedy, *$1.4bn investment in blockchain start-ups in last 9 months, says PwC expert*, 2016. Available online: http://linkis.com/Ayjzj (Accessed 1 February 2018).

11 I. Eyal, A. E. Gencer, E. G. Sirer, R. Van Renesse, *Bitcoin-NG: A scalable blockchain protocol*, in: *13th USENIX Symposium on Networked Systems Design and Implementation (NSDI 16)*, Santa Clara, CA, USA, 2016, pp. 45–59.

12 Litecoin, 2011. https://litecoin.org (Accessed 4 February 2018).

13 Y. Sompolinsky, A. Zohar, Accelerating bitcoin's transaction processing, in: Fast Money Grows on Trees, Not Chains. IACR Cryptology EPrint Archive, vol. 881, 2013.

14 C. Decker, R. Wattenhofer, *A fast and scalable payment network with bitcoin duplex micropayment channels*, in: *Symposium on Self-Stabilizing Systems*, Edmonton, AB, Canada, Springer, 2015, pp. 3–18.

15 S. Dutta, K. Saini, Statistical assessment of hybrid Blockchain for SME sector, *WSEAS Trans. Syst. Control* 16 2021. E-ISSN: 2224-2856. doi:10.37394/23203.2021.16.6.

16 N. Narayanan, K. P. Arjun, K. Saini, A Blockchain technology for asset management in multinational operation, *Essential Enterprise Blockchain Technology and Applications*, to be published by CRC Press Taylor & Francis, 2021, pp. 153–178.

17 P. Raj, K. Saini, C. Surianarayanan, (2020). *Blockchain technology and applications* (1st ed.). CRC Press. doi: 10.1201/9781003081487, ISBN-10: 0367533405, ISBN-13: 978-0367533403.

18 X. Li, P. Jiang, T. Chen, X. Luo, Q. Wen, A survey on the security of blockchain systems, *Future Generation Computer Systems*. 2017, 107, 841–853.

19 I. Eyal, E. G. Sirer, *Majority is not enough: bitcoin mining is vulnerable*, in: *International Conference on Financial Cryptography and Data Security*, San Juan, Puerto Rico, Springer, 2014, pp. 436–454.

20 J. Bonneau, E. W. Felten, S. Goldfeder, J. A. Kroll, A. Narayanan, Why Buy when You Can Rent? Bribery Attacks on Bitcoin Consensus, Citeseer, 2016.

21 G. Karame, E. Androulaki, S. Capkun, Two bitcoins at the price of one? Doublespending attacks on fast payments in bitcoin, *IACR Cryptology ePrint Archive* 2012 (248) 2012.

22 S. Dutta, K. Saini. Blockchain and social media. *Blockchain Technology and Applications*. Auerbach Publications, 2020, pp. 101–114.

23 K. Saini, Chapter Title Blockchain foundation, *Essential Enterprise Blockchain Technology and Applications*, CRC Press, 2021.

24 K. Saini (Ed.), P. R. Chelliah (Ed.), D. K. Saini (Ed.), *Essential Enterprise Blockchain Concepts and Applications*. Auerbach Publications. ISBN 9780367564889, 2021

25 *SegWit2x backers cancel plans for bitcoin hard fork*, 2017. Available online: https://techcrunch.com/2017/11/08/segwit2x-backers-cancel-plansfor-bitcoin-hard-fork/ (Accessed 1 February 2018).

26 E. B. Sasson, A. Chiesa, C. Garman, M. Green, I. Miers, E. Tromer, M. Virza, *Zerocash: decentralized anonymous payments from bitcoin*, in: *Security and Privacy (SP), 2014 IEEE* Symposium on, San Jose, CA, USA, IEEE, 2014, pp. 459–474.

27 I. Miers, C. Garman, M. Green, A. D. Rubin, *Zerocoin: anonymous distributed e-cash from bitcoin*, in: *Security and Privacy (SP), 2013 IEEE Symposium* on, Berkeley, CA, USA, IEEE, 2013, pp. 397–411.

28 Monero, 2017. Available online: https://getmonero.org/. (Accessed 20 October 2017).

29 Bitcoin Fog, 2017. Available online: http://bitcoinfog.info/. (Accessed 1 February 2018).

30 G. Maxwell, *CoinJoin: bitcoin privacy for the real world, in: Post on Bitcoin Forum*, 2013 Available online: https://bitcointalk.org/index.php?topic=279249.msg2983902 #msg2983902 (Accessed 1 February 2018).

31 A. Greenberg, 'Dark Wallet' is about to make Bitcoin money laundering easier than ever, 2014. Available online: http://www.wired.com/2014/04/dark-wallet

32 Dash, 2017. Available online: https://www.dash.org/es/. (Accessed 20 October 2017).

33 J. Bonneau, A. Narayanan, A. Miller, J. Clark, J.A. Kroll, E.W. Felten, *Mixcoin: anonymity for bitcoin with accountable mixes, in: International Conference on Financial Cryptography and Data Security*, San Juan, Puerto Rico, Springer, 2014, pp. 486–504.

34 T. Ruffing, P. Moreno-Sanchez, A. Kate, *Coinshuffle: practical decentralized coin mixing for bitcoin, in: European Symposium on Research in Computer Security*, Heraklion, Crete, Greece, Springer, 2014, pp. 345–364.

35 G. Maxwell, CoinSwap: Transaction graph disjoint trust less trading, CoinSwap: Transaction graph disjoint trust less trading (October 2013), 2013.

36 L. Valenta, B. Rowan, *Blindcoin: blinded, accountable mixes for bitcoin, in: International Conference on Financial Cryptography and Data Security*, San Juan, Puerto Rico, Springer, 2015, pp. 112–126.

37 A. Kosba, A. Miller, E. Shi, Z. Wen, C. Papamanthou, *Hawk: the blockchain model of cryptography and privacy-preserving smart contracts*, in: *Security and Privacy (SP), 2016 IEEE Symposium* on, San Jose, CA, USA, IEEE, 2016, pp. 839–858.

38 G. Zyskind, O. Nathan, A. Pentland, Enigma: Decentralized computation platform with guaranteed privacy, 2015, arXiv preprint arXiv:1506.03471.

39 Quorum Whitepaper, 2016. Available online: https://github.com/jpmorganchase/quorum-docs/blob/master/Quorum%20Whitepaper%20v0.1.pdf (Accessed 1 February 2018).

40 G. Greenspan, *MultiChain Private BlockchainWhite Paper*, 2015. Available online: https://www.multichain.com/download/MultiChain-White-Paper.pdf (Accessed 1 February 2018).

41 S. Jehan, *Rockchain A distributed data intelligence platform*, 2017. https://icobazaar. com/static/4dd610d6601de7fe70eb5590b78ed7cd/RockchainWhitePaper.pdf. (Accessed 20 October 2017).

42 A. Lazarovich, Invisible Ink: Blockchain for Data Privacy (Ph.D. thesis), Massachusetts Institute of Technology, 2015.

43 G. Zyskind, O. Nathan, et al., *Decentralizing privacy: using blockchain to protect personal data*, in: *Security and Privacy Workshops (SPW), 2015 IEEE*, San Jose, CA, USA, IEEE, 2015, pp. 180–184.

44 D. Houlding, Healthcare Blockchain: What Goes On Chain Stays on Chain, 2017. Available online: https://itpeernetwork.intel.com/healthcare-blockchain-goes-chain-stays-chain/ (Accessed 1 February 2018).

45 F. Zhang, E. Cecchetti, K. Croman, A. Juels, E. Shi, *Town crier: an authenticated data feed for smart contracts*, in: *Proceedings of the 2016 ACM SIGSAC Conference on Computer and Communications Security*, Vienna, Austria, ACM, 2016, pp. 270–282.

46 K. Delmolino, M. Arnett, A. Kosba, A. Miller, E. Shi, *Step by step towards creating a safe smart contract: lessons and insights from a cryptocurrency lab*, in: *International Conference on Financial Cryptography and Data Security*, Christ Church, Barbados, Springer, 2016, pp. 79–94.

47 N. Atzei, M. Bartoletti, T. Cimoli, *A survey of attacks on ethereum smart contracts (sok)*, in: *International Conference on Principles of Security and Trust*, Uppsala, Sweden, Springer, 2017, pp. 164–186.

48 K. Christidis, M. Devetsikiotis, Blockchains and smart contracts for the internet of things, *IEEE Access* 4 (2016) 2292–2303.

49 L. Luu, D.-H. Chu, H. Olickel, P. Saxena, A. Hobor, *Making smart contracts smarter*, in: *Proceedings of the 2016 ACM SIGSAC Conference on Computer and Communications Security*, Vienna, Austria, ACM, 2016, pp. 254–269.

50 C.K. Frantz, M. Nowostawski, From *institutions to code: towards automated generation of smart contracts*, in: *Foundations and Applications of Self Systems, IEEE International Workshops* on, Augsburg, Germany, IEEE, 2016, pp. 210–215.

51 C.K. Elwell, M.M. Murphy, M.V. Seitzinger, Bitcoin: questions, answers, and analysis of legal issues, Congressional Research Service, 2013. Available online: https://fas.org/sgp/crs/misc/R43339.pdf (Accessed 1 February 2018).

52 Bitcoin is a fraud that will blow up, says JP Morgan boss, 2017. Available online: https://www.theguardian.com/technology/2017/sep/13/bitcoin-fraudjp-morgan-cryptocurrency-drug-dealers (Accessed 1 February 2018).

53 Bitcoin could be here for 100 years but it's more likely to 'totally collapse', Nobel laureate says, 2018. Available online: https://www.cnbc.com/2018/01/19/bitcoin-likely-to-totally-collapse-nobel-laureate-robert-shiller-says.html. (Accessed 1 February 2018).

54 Bitcoin could hit $100,000 in 10 years, says the analyst who correctly called its $2,000 price, 2017. Available online: https://www.cnbc.com/2017/05/31/bitcoin-price-forecast-hit-100000-in-10-years.html (Accessed 1 February 2018).

55 E.B. Centralny, Virtual currency schemes—a further analysis, Luty, 2015. Available online: https://www.ecb.europa.eu/pub/pdf/other/virtualcurrencyschemesen.pdf. (Accessed 1 February 2018).

56 BitLegal, 2017. Available online: http://bitlegal.io/. (Accessed 1 February 2018).

57 Regulatory fears hammer bitcoin below $10,000, half its peak, 2017. Available online: https://www.reuters.com/article/uk-global-bitcoin/regulatoryfears-hammer-bitcoin-below-10000-half-its-peak-idUSKBN1F60CG. (Accessed 1 February 2018).

58 R3, 2017. Available online: https://www.r3.com/. (Accessed 1 February 2018).

59 Trusted IoT Alliance, 2017. Available online: https://www.trusted-iot.org/. (Accessed 1 February 2018).

60 Alastria: National Blockchain Ecosystem, 2017. Available online: https://alastria.io/. (Accessed 1 February 2018).

61 C. Cachin, M. Vukolić, Blockchains Consensus Protocols in the Wild, 2017, arXiv preprint arXiv:1707.01873.

62 A. Baliga, Understanding Blockchain Consensus Models, 2017. Available online: https://www.persistent.com/wp-content/uploads/2017/04/WP-Understanding-Blockchain-Consensus-Models.pdf. (Accessed 4 April 2018).

63 F. Tschorsch, B. Scheuermann, Bitcoin and beyond: a technical survey on decentralized digital currencies, *IEEE Communications Surveys & Tutorials* 18 (3) (2016) 2084–2123.

64 N.T. Courtois, On the longest chain rule and programmed self-destruction of cryptocurrencies, 2014, arXiv preprint arXiv:1405.0534.

65 Nxt White Paper, 2014. Available online: https://bravenewcoin.com/assets/Whitepapers/NxtWhitepaper-v122-rev4.pdf. (Accessed 2018-03-04).

66 F. Schuh, D. Larimer, Bitshares 2.0: General overview, 2017. Available online: https://bravenewcoin.com/assets/Whitepapers/bitshares-general.pdf. (Accessed 4 March 2018).

67 I. Stewart, Proof of burn. bitcoin. it, 2012. Available online: https://en.bitcoin.it/wiki/Proof_of_burn (Accessed 4 March 2018).

68 A. Nember, NEM Technical Reference, 2018. Available online: https://nem.io/wp-content/themes/nem/files/NEM_techRef.pdf. (Accessed 4 March 2018).

69   A. Miller, A. Juels, E. Shi, B. Parno, J. Katz, *Permacoin: repurposing bitcoin work for data preservation*, in: *Security and Privacy (SP), 2014 IEEE Symposium* on, San Jose, CA, USA, IEEE, 2014, pp. 475–490.

70   L. Lamport, et al., Paxos made simple, *ACM Sigact News* 32 (4) (2001) 18–25.

71   M. Burrows, *The chubby lock service for loosely-coupled distributed systems*, in: *Proceedings of the 7th Symposium on Operating Systems Design and Implementation*, Seattle, WA, USA, USENIX Association, 2006, pp. 335–350.

72   D. Ongaro, J.K. Ousterhout, *In search of an understandable consensus algorithm.*, in: *USENIX Annual Technical Conference*, Philadelphia, PA, USA, USENIX Association, 2014, pp. 305–319.

73   M. Nabi-Abdolyousefi, M. Mesbahi, Sieve method for consensus-type network tomography, *IET Control Theory Appl.* 6 (12) (2012) 1926–1932.

74   Ripple, 2017. https://ripple.com/. (Accessed 20 October 2017).

75   D. Mazieres, *The stellar consensus protocol: a federated model for internetlevel consensus*, Stellar Development Foundation (2015).

76   HDAC, 2017. Available online: https://hdac.io/. (Accessed 1 February 2018).

77   V. Gramoli, From blockchain consensus back to byzantine consensus, *Future Gener. Comput. Syst.* 107 (2017) 760–769.

78   J.C. Buzby, T. Roberts, The economics of enteric infections: human foodborne disease costs, *Gastroenterology* 136 (6) (2009) 1851–1862.

79   H. Malviya, How Blockchain will Defend IOT, 2016. Available online: https://ssrn.com/abstract=2883711 (Accessed 1 February 2018).

80   P. Veena, S. Panikkar, S. Nair, P. Brody, *Empowering the edge-practical insights on a decentralized internet of things*, in: *Empowering the Edge Practical Insights on a Decentralized Internet of Things*, vol. 17, IBM Institute for Business Value, 2015.

81   S. Gan, *An IoT Simulator in NS3 and a Key-Based Authentication Architecture for IoT Devices using Blockchain*, Indian Institute of Technology Kanpur, 2017.

82   Chain of things, 2017. Available online: https://www.blockchainofthings.com/ (Accessed 1 February 2018).

83   Filament, 2017. Available online: https://filament.com/. (Accessed 1 February 2018).

84   Modum, 2017. Available online: https://modum.io/. (Accessed 1 February 2018).

85   G. Prisco, Slock. It to introduce smart locks linked to smart ethereum contracts, decentralize the sharing economy, 2016. Available online: https://bitcoinmagazine.com/articles/slock-it-to-introduce-smart-locks-linked-to-smart-ethereum-contracts-decentralize-the-sharing-economy-1446746719/ (Accessed 1 February 2018).

86   M.A. Khan, K. Salah, Iot security: review, blockchain solutions, and open challenges, *Future Gener. Comput. Syst.* 82 (2017) 395–411.

87   LO3ENERGY, 2017. Available online: https://lo3energy.com/. (Accessed 1 February 2018).

88   Aigang, 2017. Available online: https://aigang.network/. (Accessed 1 February 2018).

89   My bit, 2017. Available online: https://mybit.io/. (Accessed 1 February 2018).

90   M. Samaniego, R. Deters, *Hosting virtual iot resources on edge-hosts with blockchain*, in: *Computer and Information Technology (CIT), 2016 IEEE International Conference* on, Yanuca Island, Fiji, IEEE, 2016, pp. 116–119.

91   M. Aazam, E.-N. Huh, *Fog computing and smart gateway based communication for cloud of things*, in: *Proceedings of the 2nd International Conference on Future Internet of Things and Cloud, FiCloud-2014*, Barcelona, Spain, August 2014, pp. 27–29.

92   Ethembedded, 2017. Available online: http://ethembedded.com/. (Accessed 1 February 2018).

93   Raspnode, 2017. Available online: http://raspnode.com/. (Accessed 1 February 2018).

94 K. Wüst, A. Gervais, Do you need a blockchain? *IACR Cryptology EPrint Archive* 2017 (2017) 375.

95 Ant Router R1-LTC The WiFi router that mines Litecoin, 2017. Available online: https://shop.bitmain.com/antrouter_r1_ltc_wireless_router_and_asic_litecoin_miner.html (Accessed 1 February 2018).

96 Ethraspbian, 2017. Available online: http://ethraspbian.com/. (Accessed 1 February 2018).

97 R. Roman, J. Lopez, M. Mambo, Mobile edge computing, fog et al.: a survey and analysis of security threats and challenges, *Future Gener. Comput. Syst.* 78 (2018) 680–698.

98 R. Roman, J. Zhou, J. Lopez, On the features and challenges of security and privacy in distributed internet of things, *Comput. Netw.* 57 (10) (2013) 2266–2279.

99 J. Lopez, R. Rios, F. Bao, G. Wang, Evolving privacy: from sensors to the internet of things, *Future Gener. Comput. Syst.* 75 (2017) 46–57.

100 M. Banerjee, J. Lee, K.-K.R. Choo, A blockchain future to internet of things security: a position paper, *Digital Commun. Netw.* (2017). doi: 10.1016/j.dcan.2017.10.006. http://www.sciencedirect.com/science/article/pii/S2352864817302900

101 P. Ruckebusch, E. De Poorter, C. Fortuna, I. Moerman, Gitar: generic extension for internet-of-things architectures enabling dynamic updates of network and application modules, *Ad Hoc Networks* 36 (2016) 127–151.

102 A. Taherkordi, F. Loiret, R. Rouvoy, F. Eliassen, Optimizing sensor network reprogramming via in situ reconfigurable components, *ACM Transactions on Sensor Networks (TOSN)* 9 (2) (2013) 14.

103 C. Fernandez-Gago, F. Moyano, J. Lopez, Modelling trust dynamics in the internet of things, *Inform. Sci.* 396 (2017) 72–82.

104 C. Liu, R. Ranjan, C. Yang, X. Zhang, L. Wang, J. Chen, Mur-dpa: top-down levelled multi-replica merkle hash tree based secure public auditing for dynamic big data storage on cloud, *IEEE Trans. Comput.* 64 (9) (2015) 2609–2622.

105 C. Wang, Q. Wang, K. Ren, W. Lou, *Privacy-preserving public auditing for data storage security in cloud computing*, in: INFOCOM, 2010 Proceedings IEEE, San Diego, California, USA, IEEE, 2010, pp. 1–9.

106 C. Liu, C. Yang, X. Zhang, J. Chen, External integrity verification for outsourced big data in cloud and iot: a big picture, *Future Gener. Comput. Syst.* 49 (2015) 58–67.

107 Bitcoin Fog, 2016. Available online: http://www.the-blockchain.com/2016/05/01/babel-chain-machine-communication-proof-understanding-new-paper/ (Accessed 1 February 2018).

108 *I.A. Naidu R, Nestle, Unilever, Tyson and others team with IBM on blockchain, Reuters*, 2017. http://www.reuters.com/article/us-ibm-retailers-blockchain/nestle-unilever-tyson-and-others-team-with-ibm-on-blockchain-idUSKCN1B21B1 (Accessed 20 October 2017).

109 M. Samaniego, R. Deters, *Internet of smart things-iost: using blockchain and clips to make things autonomous*, in: Cognitive Computing (ICCC), 2017 IEEE International Conference on, Honolulu, Hawaii, USA, IEEE, 2017, pp. 9–16.

110 The Lisk Protocol, 2017. Available online: https://docs.lisk.io/docs/the-liskprotocol. (Accessed 1 February 2018).

111 Chronicled, 2017. Available online: https://chronicled.com/. (Accessed 1 February 2018).

112 Reyna, Ana, Cristian Martín, Jaime Chen, Enrique Soler, and Manuel Díaz. "On blockchain and its integration with IoT. Challenges and opportunities." *Future generation computer systems* 88 (2018): 173–190.

12 Blockchain for Medicine and Healthcare

Mohit Dayal, Rupesh Kumar Garg, Sweta Chaudhary, Chitrangada Chaubey, and Ameya Chawla

CONTENTS

12.1 INTRODUCTION

The internet nowadays has great importance for all of us to connect billions of people around the world. It is used for communicating and collaborating online. However, it was initially built for moving and storing information. It has changed the way we do business. Now, for the first time in history, two or more parties anywhere in the world can transact and do their business peer to peer with the help of blockchain. Nowadays, blockchain – "the trust protocol" – represents the second era of the internet.

The internet was initially designed to move information. It was not created for the purpose of sending or managing assets.

There are two types of systems available:

1. Centralized
2. Decentralized

In a centralized system, there is a single authority and everyone in the system is relying on the single authority for availability of data. In our day-to-day life, most systems are centralized in nature. Examples are Facebook, Google, Uber, Amazon, Netflix, etc. Centralized systems also have the drawbacks, such as:

1. If the central authority fails, our whole system is not working.
2. It is also possible that central authority is biased to some participants in the system.

Unlike the centralized system, a decentralized system doesn't have a single authority. In fact, all the participants in the system have a copy of the data. It means if anyone can change the database, others can also see their reflection.

To overcome the drawbacks of centralized systems, blockchain was developed. It was first introduced by researchers in 1991. But it was mainly used in bitcoin in 2009. Nowadays, bitcoin application has a very high demand in cryptocurrencies.

12.1.1 WHAT IS BLOCKCHAIN?

It is a platform for exchanging digital assets peer to peer, from me to you. That's why we call it an internet of value.

Blockchain is defined as a decentralized distributed ledger that records the provenance of exchanging a digital asset. Ledger refers to the immutable database [1].

In other words, it has a chain of blocks connected to each other with the help of previous block hash. Each block contain three things: hash, data, previous block hash (Figure 12.1).

The above figure says that each block is connected to previous block hash and a new hash for that particular block hash is calculated by combining data and previous block hash.

12.1.2 BLOCKCHAIN IS TRUSTWORTHY

Suppose we change the data of second block (Figure 12.2).

Then it will change the hash of that particular block. Then ultimately it will change every block hash afterwards (Figure 12.3).

Then the hash of the particular block mismatch with the next block. As a result it will inform that data is change in the system (Figures 12.4 and 12.5).

Now the next block hash also changes accordingly (Figures 12.6 and 12.7).

Hence, we will see that if we change the data in our system, it will change the subsequent blocks afterwards. Hence, it is trustworthy.

- Linked list
  1. Replicated
  2. Distributed

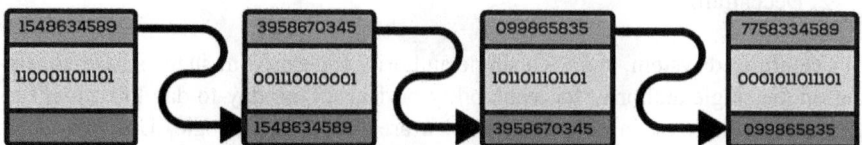

FIGURE 12.1 Chains of blocks.

3. Consistency maintained by consensus
4. Cryptographically linked
5. Cryptographically assured integrity of data

- Used as:
  1. Immutable ledger of events, transactions, or time-stamped data
  2. Tamper-resistant log
  3. Platform to create and transact in cryptocurrencies
  4. Log of events/transactions unrelated to currency.

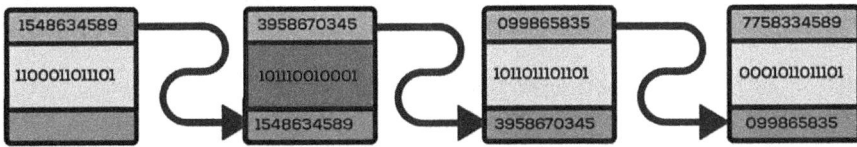

FIGURE 12.2   Second block data changes.

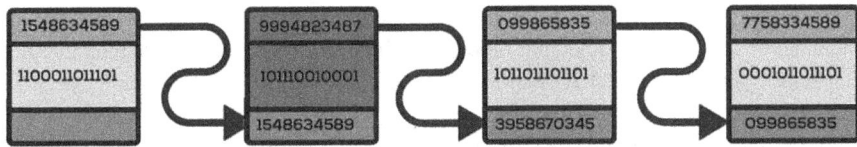

FIGURE 12.3   Second block hash changes.

FIGURE 12.4   Mismatch with the next block.

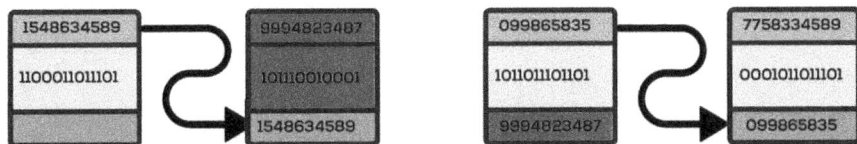

FIGURE 12.5   Third block previous hash changes.

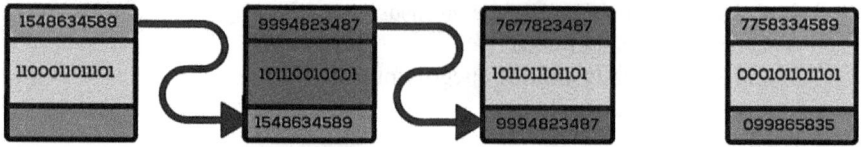

FIGURE 12.6 Third block hash changes.

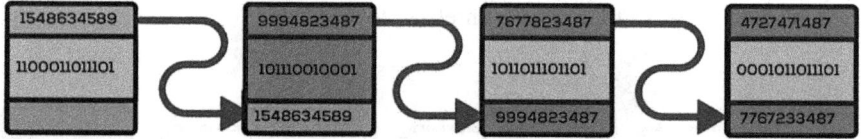

FIGURE 12.7 Similarly with fourth block.

We all know that a linked list is a set of blocks which are connected to each other by some kind of a link. In case of a data structure linked list, you have the nodes, and nodes are connected by pointers and pointers are basically memory addresses. But, in case of a blockchain technology, the blockchain is a distributed data structure, that is replicated at various nodes or various computers and therefore, the linking is not based on memory addresses. So we have a different notion of linking between nodes and each of these nodes are called blocks. So therefore, we can imagine a blockchain as a series of blocks and each block is connected to its previous block and later block by some kind of a link.

And so it is replicated all over because replication gives you a number of different advantages like if one of the replica gets corrupted, the other replicas are there to make sure that the integrity of the information contained in the in the data structure is maintained. And also replication gives you some kind of guarantee of integrity of data. Therefore, blockchain is a data structure that is maintained distributedly and that is replicated and main aim of blockchain is to maintain the full integrity of the data.

And what is the integrity? Integrity means that the data once it has been agreed by all the relevant parties to put in the data structure, it has not been tampered with.

Nobody has come and changed the data and claimed that this is the data that was put in. The blockchain is designed in such a way, so that it is an immutable ledger of events, which means it is a log that cannot be changed by a malicious party or by mistake. And therefore, all the data that you put in there could be event logs, it could be transactions, it could be various kinds of accesses and modifications you do to some other thing like a data or you do a property transaction.

When you make a banking transaction, the bank keeps logs of when you interacted with its banking servers and what you did, what transactions you made; all this is logged. The main problem with keeping logs without any notion of protection of the integrity is that somebody can tamper with the logs and somebody can delete some of the accesses. And therefore, later on when you check the log, you would know some part of its history. All these things logs are has to be kept in an immutable ledger. And that is what blockchain provides.

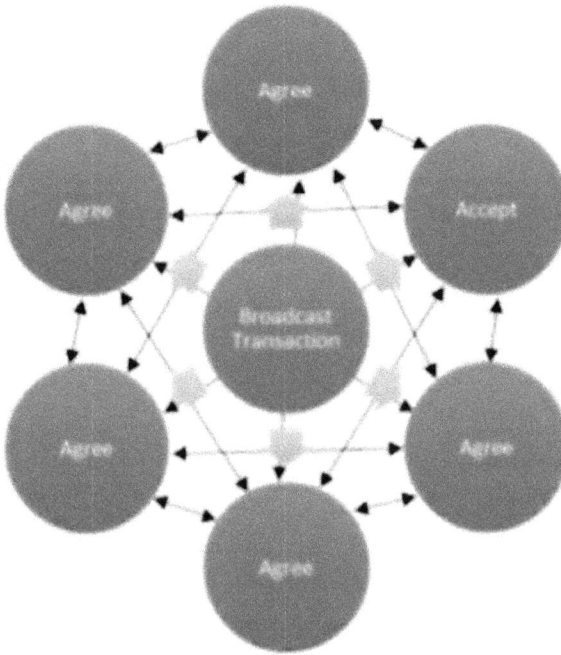

FIGURE 12.8 Decentralized consensus between participants.

The internet of things on the other hand also play a main role in a network of connected devices. The goal of IoT is to provide a reliable and protected exchange of things, but the security challenges will be solved by cryptographic techniques. These techniques include user authentication and knowledge privacy [2].

12.2 THE NAKAMOTO CONSENSUS MECHANISM

Consensus as an emergent phenomenon

The main invention by the creator of bitcoin Satoshi Nakamoto, is a decentralized consensus mechanism. It gives us provision for the participants to agree on the contents of the database (Figure 12.8).

If we want to change our data then there is a majority of voting in favor of changing the data. Only then can we change the data [3].

There is no single point of failure in the blockchain as it is decentralized and replicated.

12.3 PEER-TO-PEER NETWORKS

The data in the blockchain is connected peer to peer. It means the data in one system has the same data in other system also. Each node in the system is connected with other nodes in the system. It means if we want to change the data in one computer

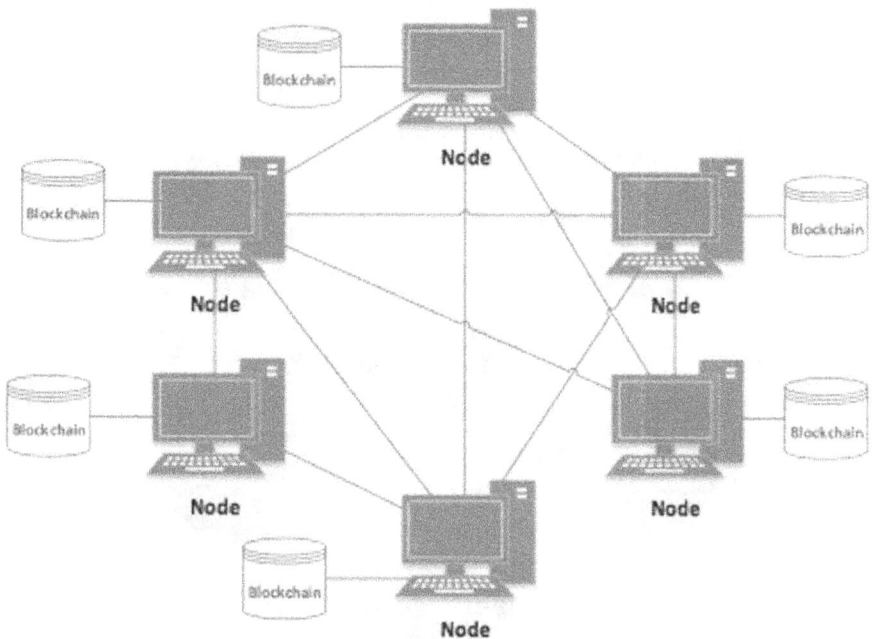

FIGURE 12.9 Peer-to-peer network.

then we should also change the data in other computer in system connected in the peer-to-peer manner (Figure 12.9).

12.4 HOW WE ADD A NEW BLOCK IN OUR SYSTEM

The process given by Satoshi Nakamoto in 2008 is explained in following steps:

1. When a new transaction comes, it will first be broadcast to all nodes.
2. In the second step, each and every node collects this new transaction into a particular block.
3. In the next step, each node is working on finding a consensus for this block.
4. In the next step, when a node gets a consensus it will broadcasts the block to all available nodes.
5. Now, if all the transactions in the block are marked as valid and not behave malicious, then only nodes will accept the block.
6. Lastly, nodes express their acceptance of the block by working on creation of the next block in the chain, by joining the hash of the accepted block as the previous hash [3].

Now, we come to the point that the data in our system is transparent and immutable.

12.5   APPLICATIONS

Now we come to say that we can use blockchain in such a case where a constant amount of data is required, so that no one can change it.

1. In real estate for storing information related to house owner.
2. Online voting
3. Digital Identity card for storing information such as birth certificate, unique voter ID card, passport, pan card, Aadhar card, driving license, medical records, insurance, etc.
4. In banking and finance
5. In medicine and healthcare
6. In Internet of Things
7. In travel and transportation
8. In cryptocurrencies like bitcoin, etc.

Now we will discuss blockchain in medicine and healthcare in detail. One of the best uses of blockchain is in healthcare.

Developed and even developing countries invest a large amount of their gross domestic product (GDP) on the healthcare domain. In the meantime, hospital costs are too expensive, along with unfit ways and health data losses are poised to continue.

This is one such area in which we can use blockchain. It can do for us many things such as storing guarded patient information to handling the epidemics.

Main concerns related with blockchain applications in healthcare include:

1. A large network infrastructure at different levels.
2. There should be Identity verification and authorization of all participants.
3. There should be constant patterns of verification process to access electronic health data or information.

There should be a distributed data of patients that will be shared between different hospitals but there must be a power given to patients to share their data.

12.6   HOW BLOCKCHAIN WILL CHANGE THE HEALTHCARE SYSTEM

a) **Health records**:
   We know very well that electronic medical records (EMR) are the pillar or foundation of our modern healthcare system. Each time we visit the doctor, these medical records grow. And as every hospital and every doctor's office has its own way of storing the patient information, it is always difficult for healthcare providers to obtain them. The basic purpose of many companies like Patientory, Medibloc, and Medicalchain is to solve this problem. The purpose of this is to give patients authentication over their entire medical

history and to provide a flexible access to it for patients and physicians as well. Blockchain as a combine would simplify the data access more efficient, and also bring data security to the field as well [4].

As the data grows day by day, the medical records are very complex and it is very hard to process this complex data using traditional tools. We can here also use big data analytics to process raw data into meaningful information so that we can use it [5].

Big data means a lot of data. This huge amount of medical data is used for data analysis and research purposes. Big organizations use this big data to predict the future event occurrence such as the number of patients having the same kind of diseases in a particular time period in last year. We can use big data for analysis and then forming some required information related to it [6].

b) **Supply chains**:
In recent times, the pharmaceutical industry came up with the biggest standards for product cover, security, and firmness of product. As an example, if the supply chain management system is working in combination with blockchain it can be monitored, protected and transparently. It can largely decrease time delays and human mistakes and errors. It should also be used to monitor costs, labor, and even unwanted material that is released at every stages in the supply chain. It should also be used to give confirmation about the verification of products by following them from the starting point, that makes the drug and medicine market more efficient and prevent them from losses annually. Many companies such as Chronicled, Blockpharma and Modum are so far working toward the aim of more structured blockchain logistic solutions. Companies such as Modum specially works in the domain of consent with EU laws which require evidence that medicinal and drug related products should not been reveal to special conditions, especially at certain temperatures (high temperatures), that will cover up the quality of their products. The solution of Modum was to build a sensor which records and track the environmental conditions. But in the mean while physical products are in transportation and permanently record and save this data on the blockchain [7].

c) **Genomic market**:
Many companies like EncrypGen and Nebula Genomics are already working on it to build blockchain platforms to authorize people to share their genomic data safety and highly secured on a new growing market. They gamble that in the future, the chances around the personal genome sequencing will create a big data market in very high cost. The best technology to solve these problems and to ensure that data gets from the origin to its end-users is no one the blockchain. These companies are highly focused and their aim is to use blockchain technology to manage genomic data

protection efficiently; this will enable buyers to properly acquire genomic data and manage the challenges behind it. These companies are new in the market, called startups [8].

The advantages of applying blockchains, in comparison with old ways of storing healthcare database systems that were comprised of fixed databases, are data decentralization, data origin, manage data, vigorous data and availability of data to the verified users when it is required.

The database should be kept secure. Nowadays cyber security targets the data and sometimes the unknowingly attack. This will also include the cyber criminals and cyber risks. So, to prevent cyber-attacks, we have to use the risk management with security architecture. We have to take countermeasures or preventions to get security [9].

It can be applied in five primary areas:

1. It gives us a way to manage electronic medical records (EMR).
2. It provides shielding of healthcare data.
3. It provides a way for everyone to check its own health record data management.
4. It gives us a way for point-of-care genomics management.
5. It gives electronic health records data management.

12.7 SOME SPECIFIC APPLICATIONS

a) **Research**:

Nowadays, the automatic modifying and sharing the medical information of a given patient within an organization is allowed by the electronic health records. This set of data will allow the researchers and other organizations to get this large amount of data. As a result, this greatly influences clinical research, identification of health reporting, creation of public health reporting and safety events and its unfavorable effect reporting.

b) **Seamless movement of patients' data among different providers**:

Data **stored** in the blockchain allow each and every individual patients to easily verify and share their health data with other providers or any organizations with the help of shareable private key. It should create health information technology(HIT) more efficient and sharable among different users [10].

c) **Quick, economical and better patient care**:

It has a one system for storage, constantly modifying health data for faster and safely access by verified users. Miscommunication among different healthcare professionals or doctors caring for the same patient must be prevented. These errors can be avoided and as a result faster diagnosis and proper care become possible for each patient.

d) **Data security**:
The blockchain provides a high data security by providing each individual a key or public identifier and private key which can be unlocked for a particular duration. Blockchain also provides immutable ledger of events. As a result, our health information is stored more efficiently and in a secure manner.

e) **Mobile health applications and remote tracing**:
Nowadays, mobile health apps are becoming more popular with advance technology. In this regard, electronic medical records are safely stored and easily accessible by medical personnel as and when required. As a result blockchain comes into picture [11].

f) **Monitoring and safety of medical supplies**:
It can help to secure and identify the supply of pharmaceuticals with full transparency. It is also providing the tracing of the labor costs and carbon emissions involved in the formation of these supplies.

g) **Health insurance assert**:
Blockchain is used to store the information related to the healthcare claims as it is immutable in nature. As a result, no one can change it. It should be transparent to the users.

h) **Tracking diseases and outbreaks**:
It helps us to recognize real-time diseases and then report and explore the disease patterns which can help us to identify their origin and transmission parameters [12].

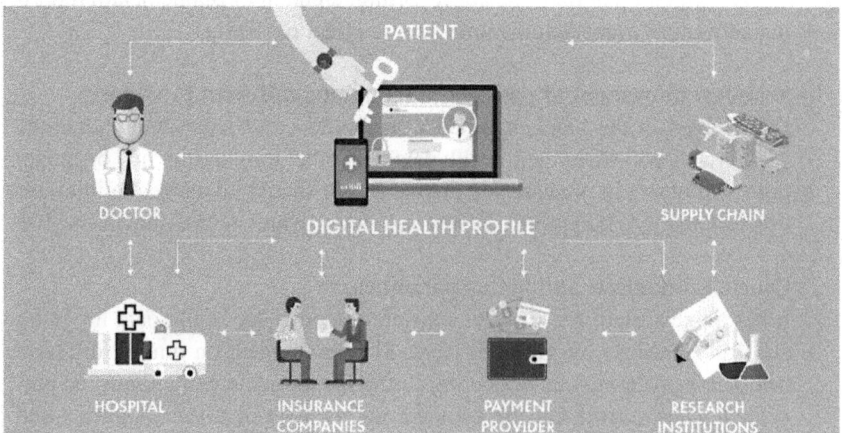

FIGURE 12.10   Use cases in medicine and healthcare.

At last, we come to conclude that blockchain has become a very big role if it is used in efficient manner specially in healthcare. As a result, the cost is reduced and proper healthcare systems should be maintained where the data is secured and distributed (Figure 12.10).

12.8 FUTURE OPPORTUNITIES

There are many open challenges in the field of blockchain in terms of security as well as to reduce the complexity. Some advanced techniques are also taken into action to make the blockchain system more reliable in terms of faster execution. There is still much research going on to tackle the drawbacks [10].

REFERENCES

1 G. Drosatos, "Blockchain applications in the biomedical domain: A scoping review," *Biotechnology Journal*, 2019. [Online]. Available: https://doi.org/10.1016/j.csbj. 2019.01.010. https://www.sciencedirect.com/science/article/pii/S200103701830285X.

2 M. Khari, A. K. Garg, A. H. Gandomi, R. Gupta, R. Patan and B. Balusamy, "Securing data in Internet of Things(IOT) using cryptography and stenography techniques," *IEEE Transactions on Systems,Man, and Cybernetics: Systems 50* (1), pp. 73–80, 2019.

3 S. Nakamoto, "Bitcoin: A peer-to-peer electronic cash system," 2008. [Online]. Available: https://bitcoin.org/bitcoin.pdf.

4 S. Khezr, "Blockchain technology in healthcare: a comprehensive review and directions for future research," *Applied Sciences*, 2019. [Online]. Available: https://doi.org/10.3390/ app9091736. https://www.mdpi.com/2076-3417/9/9/1736/htm.

5 M. Dayal and N. Singh, "An anatomization of Aadhar card data set – A big data challenge," *Procedia Computer Science 85*, pp. 733–739, New Delhi, 2016.

6 M. Dayal and N. Singh, "Indian health care analysis using big data programming tool," *Procedia Computer Science* 89, pp. 521–527, New Delhi, 2016.

7 R. Ben Fekih and M. Lahami, "Application of blockchain technology in healthcare: A comprehensive study. The impact of digital technologies on public health in developed and developing countries," 2020. [Online]. Available: https://www.ncbi.nlm.nih.gov/ pmc/articles/PMC7313278/.

8 S. Daley, "How using blockchain in health care is reviving the industry's capabilities," 1 July 2019. [Online]. Available: https://builtin.com/blockchain/blockchain-healthcare-applications-companies.

9 M. Khari, G. Shrivastava, S. Gupta and R. Gupta, "Role of cyber security in today's scenario," in *In Detecting and Mitigating Robotic Cyber Security Risks*, 2017, IGI Global.

10 www2.deloitte.com, "Blockchain: Opportunities for health care," 2020. [Online]. Available: https://www2.deloitte.com/us/en/pages/public-sector/articles/blockchain-opportunities-for-health-care.html.

11 H. S. chen, "Blockchain in healthcare: A patient-centered model," *Biomedical Journal of Scientific & Technical Research*. doi:10.26717/BJSTR.2019.20.003448, 2019. [Online]. Available: https://biomedres.us/fulltexts/BJSTR.MS.ID.003448.php.

12 L. Thomas, "Blockchain Applications in health care," 19 January 2021. [Online]. Available: https://www.news-medical.net/health/Blockchain-Applications-in-Healthcare.aspx.

Index

For Product Safety Concerns and Information please contact our EU
representative GPSR@taylorandfrancis.com
Taylor & Francis Verlag GmbH, Kaufingerstraße 24, 80331 München, Germany